Technical Reporting

Technical Reporting

THIRD EDITION

JOSEPH N. ULMAN, JR.
Consumer Reports *Magazine*

JAY R. GOULD
Rensselaer Polytechnic Institute

HOLT, RINEHART AND WINSTON, INC.
New York Chicago San Francisco Atlanta
Dallas Montreal Toronto London Sydney

Library of Congress Catalog Card Number: 70–158149
ISBN: 0–03–081003–5
Printed in the United States of America
6 038 9 8 7 6 5 4

Preface
to the Third Edition

Audience

This book is addressed primarily to students and practitioners of engineering and the sciences who have reached the point at which they have reporting jobs to do and have something to say. Most technical students attain this stage in their sophomore, junior, or senior year; a very few, in their freshman year; and some, when they are in graduate school or in industry. Only at this stage are most of them receptive to instruction in technical reporting, and only then do they have genuinely realistic subject matter to work with.

Content

The first edition of *Technical Reporting* was based on material used at the Massachusetts Institute of Technology for instruction that fell into two main categories: (1) a course in technical writing offered by the Department of Humanities and (2) instruction given in connection with reports required of the students by several of the professional departments (Mechanical, Civil, Electrical, and Marine Engineering, Physics, Geology, and others). The first revision and enlargement stemmed from experience at Rensselaer Polytechnic Institute (RPI), both in undergraduate courses and the Graduate Program in Technical Writing, and in the Technical Writers' Institute held each summer for professional technical writers and supervisors.

This third edition contains material that has come not only from technical writing courses but also from the authors' experiences as consultants in industrial concerns and agencies, as directors of in-plant training programs, and as participants in the activities of several professional writing societies.

A complete chapter on proposals has been added, together with specimens of this type of technical reporting.

The chapter on technical papers and articles has been expanded by the addition of new illustrative material, and a specimen of a student article has been placed in the Appendix.

Specimens of technical description have also been added to the Appendix.

The chapter on oral reports now includes a section on the techniques of the presentation, which we think will be valuable to executives as well as to students.

In addition, the references in Part II and the examples in Part III have been made contemporary, and many new examples have been added. Most of the illustrative specimen reports and letters in the Appendix have been replaced by new ones reflecting current state-of-the-art, and the Bibliography has been enlarged and brought up-to-date.

Finally, additional written exercises have been included at the ends of a number of chapters. We feel that the most realistic and fruitful writing assignments, with a maximum of motivation for the student, are likely to originate in the professional or technical departments, in connection with technical or experimental investigations being performed by the students. For those institutions, however, where instruction in technical writing is given by members of the English Department independently of their colleagues in the technical fields, these exercises should prove useful.

In *Technical Reporting* we have stressed the thesis that one of the commonest and most serious faults in technical and scientific writing is the burial of the important, fundamental ideas under a mass of detail—the masking of the forest by the trees. Many of the textbooks on technical writing seem to suffer from this very fault.

Accordingly, *Technical Reporting* stresses the principles—some of them quite elementary—that seem to be most overlooked by technical writers. This policy has permitted us to treat fully those subjects that are included.

For instance, we have not presented the section on old-fashioned business-letter clichés that appears in many textbooks in this field. Practically no writer under 75 ever indulges in the "Yours of the 20th inst. to hand . . ." pattern, and there seems no reason to burden a younger generation with warnings against it. On the other hand, we have devoted a lot of attention to some principles for making graphs and curves clear and easy to read—some principles that are commonly overlooked by practicing professional men as well as students.

Another valid reason for omitting a subject is the strong likelihood that most students will be unable or unwilling to remember it or use it. For example, the usual handbook on composition lists a dozen or more uses of the comma, most of them described in technical grammatical terms and given without reason or justification. The average person just doesn't remember all these formal rules; indeed he probably never understands many of them. Consequently he dismisses from his mind all he has been told about the comma, and punctuates improperly and confusingly. In the section on punctuation in this book we have suggested a few simple and easily remembered general rules of thumb for the use of the comma; but we have dwelt in detail on the three comma rules that we believe must be observed rigorously and consistently in order to eliminate ambiguity. Furthermore, we have explained *why* each rule is used, because most people neither remember unjustified rules nor employ them intelligently.

Acknowledgments

For his helpful criticism and advice we are indebted to Professor Lynwood S. Bryant of MIT; and, for his assistance in preparing some of the drawings, to Mr. Chauncey W. Watt. Dr. Sterling P. Olmsted, dean of the faculty at Wilmington College, and Professor Douglas H. Washburn of RPI have permitted us to use material previously compiled by them.

Thanks are also due to the Hercules Company for permission to reproduce the report outline from *Hercules Technical Reports—A Guide to Their Preparation;* to Chrysler Corporation for permission to reproduce the report outline from their *Technical Report Manual;* to the National Aeronautics and Space Administration for permission to reproduce portions of the report *Differential Thresholds for Motion in the Periphery,* by James M. Link and Leroy L. Vallerie; to Arthur D. Little, Inc., for permission to quote part of "The Turbo-Encabulator"; to *Research / Development,* F. D. Thompson Publications, Inc., for permission to quote from a column by Dr. Merritt Williamson; to the Friden Division of the Singer Company for permission to reproduce a section of the Technical Manual for 2201 FLEXO-WRITER automatic writing machine; to *Science Illustrated,* April 1946, for permission to reproduce $E = MC^2$, by Dr. Albert Einstein; and to Heath Company (Benton Harbor, Michigan), for permission to reproduce the diagrams of Figs. 12, 13, and 14.

We are indebted to various RPI groups and personnel for permission to use material, particularly to the following: the Research Division, Dr. Joseph V. Foa, and Dr. Henrik J. Hagerup for portions of the proposal *Feasibility of a New Mode of Mass Transportation;* Mr. Franz J. Buschmann, a former student, for portions of the report *Density-Structure Relationships in Pyrolytic Graphite;* and Mrs. Mary Albert Porter, also a former student, for the article *Chemical Lasers.*

Bethany, Connecticut
Troy, New York
October 1971

J. N. U., Jr.
J. R. G.

Abbreviated Table of Contents

Contents

I. Basic Issues

III. Tools and Methods

Technical Reporting

I

Basic Issues

1. A Selling Job

We may as well face the fact: if you are a typical undergraduate, you probably regard the writing of reports as a dull and superfluous chore.

This attitude has been underscored by Dr. Merritt A. Williamson, Dean of Engineering at Pennsylvania State University. In a magazine article, Dr. Williamson said, "Writing reports has often been the bane of the technical man's existence. I am inclined to agree, based on my experience with technical men, that there are few things which they like to do less than write reports. . . . I have a feeling that the primary difficulty in writing reports or papers is in the organization that has to precede any creative literary effort."

Perhaps another reason for this state of affairs is that the undergraduate —particularly in his earlier years—seems to have very little to say. As he progresses through college and on into graduate school or industry, he develops a body of knowledge. At some time in his career he acquires some information or some ideas that he wants to pass on to others. Only then does he wish for instruction in technical reporting. But then, all too often, he has finished his schooling, and he has to get his instruction the hard way.

Thus this book, along with most others in its field, seems to be obliged to begin with a selling job.

1.1 The Importance of Technical Reporting

The complexity of an organization increases exponentially with its size. As the complexity goes up, so too does the need for written records and communications. Only through a full exchange of information can the various divisions of a large organization coordinate their efforts effectively.

But even the small organization has a vital need for accurate technical reporting. How was a special part fabricated last year? How was a test performed? What are the precautions to be observed with a seldom-used instrument? Written records furnish authoritative answers to many questions such as these, and they increase the efficiency of the organization that maintains a rigorous reporting procedure.

Some engineering and scientific organizations do nothing but investigation, testing, experimentation, or research. Their only tangible product is the report. If they are to have anything to show for their efforts, they must do a thorough job of reporting.

On the other hand, in some companies the report is only one of several kinds of writing being done, although reporting is the foundation for the other kinds. Manuals and instruction books accompany machinery when it is sold, to be given to maintenance engineers and repairmen. Proposals must be written to obtain funds from government agencies. Papers and articles by company employees appear in national journals and technical magazines. Technical literature is provided for salesmen, and career brochures for college graduates entering engineering fields.

1.2 The Importance of Proficiency in Technical Reporting

In many engineering organizations, particularly those doing experimental work or research, the young employee's chief communication with his superiors is through his written (or oral) reports. Often the supervisor has no other criterion by which to judge an employee's work.

Every technical school receives letters from important industrial firms complaining about the quality of the technical reports produced by its graduates. These young men, we are told, are admirably grounded in the basic sciences, they are intelligent, they are capable of doing excellent work. But their education has left a serious gap: they are unable to describe clearly and succinctly what they have done. This inability exists, we believe, not so much because the engineering schools fail to offer instruction in this important subject, but because the students lack sufficient motivation to apply themselves to it.

We hope that this introduction, together with the practical approach of *Technical Reporting,* will persuade you to put real effort into the study of this subject. If you do, we believe that you will find the effort to be both challenging and interesting.

2. Important Fundamental Principles

2.1 Communication

Technical reporting is a specialized branch of the field of *communication*. Communication may be broadly defined as the transmission of information or ideas from one mind to other minds. Every communication—whether it is a telegram, a novel, or an instruction book—should have (1) a specific purpose and (2) a specific audience. It should be carefully planned and constructed to fit both.

In addition, the writer or speaker should know the most suitable format in which to put his message and the best organization for arranging his material to achieve clarity, emphasis, and consistency.

2.2 Specific Purpose

Some forms of communication may have as their primary aim to convey a feeling or an atmosphere, or to amuse, or even to stupefy or confuse. But every technical communication has one certain clear purpose: *to convey information and ideas accurately and efficiently*. This objective requires that the communication be: (1) as *clear* as possible; (2) as *brief* as possible; and (3) as *easy to understand* as possible. If at the same time it can be made pleasant, or perhaps striking or forceful, all well and good. But these considerations and all others are secondary to the primary objectives of accuracy and efficiency.

Ways of achieving these ends will be the subject of a major portion of *Technical Reporting*.

2.3 Specific Audience

Any communication, if it is to be effective and efficient, must be designed for the needs and the understanding of a specific reader or group of readers. It must lie neither beyond their powers of comprehension nor so far beneath their level of competence as to bore them and thus lose them. The successful comic book is easily followed by a moron; on the other hand, the treatise on operator theory in integral equations should not discourage its readers by including an extended discussion of Euclidean geometry.

The expected audience of a technical communication determines the treatment of several factors.

2.3.1 Background Information

Are the people who will see or hear this report familiar with the field, or do they need extensive general orientation? Are they familiar with the circumstances of the present case, or do they need briefing? How much background information must you supply them? Always err on the side of giving them too much.

2.3.2 Technical Level

What is the technical training of your audience? They may be experts in your own field; they may have general scientific training but a specialization different from yours; or they may be laymen. Correspondingly, you may be able to take for granted the most advanced knowledge of the ideas and vocabulary of your own field; you may have to explain only specialized ideas or terms; or you may have to explain fully every concept that transcends the knowledge of the man on the street. Can you use differential equations in presenting your ideas, or must you restrict yourself to algebra? Or must you avoid the use of equations entirely?

Unfortunately the answers to these questions are often far from clear-cut. Many communications go to mixed audiences, perhaps widely mixed. A compromise treatment is called for, one aimed at the less-informed members of your audience rather than the better-informed.

But don't let the frequent need for compromise prevent you from always trying to determine the nature of your expected audience and fitting your presentation to them just as closely as you can.

Herbert B. Michaelson, associate editor of the *IBM Journal of Research and Development,* has written about the writer's responsibilities. In part, Mr. Michaelson says:

> *The writer shoulders the initial responsibility for making his manuscript intelligible. Wherever possible, he should anticipate what class of readers will see his report and should adopt a style of language they can understand. Particularly when writing for workers in his own field, he should use their standard terminology, if such standardization exists. If not, he might add a glossary to his report. He should be aware of the existence of multivalued words and should look for them when revising his rough draft. When describing specific details of his work, he must avoid vague, generalized expressions.*

2.4 A Framework for the Parts

A highly important principle of technical reporting has been stated so well by Professor Lynwood S. Bryant of the Massachusetts Institute of Technology that we present it in his words:

> *The first thing an observer needs when he is confronted with a strange machine (or a strange concept) is a clear idea of the essence and function and purpose of the whole, considered as a unit. He needs an intelligible whole to fit the parts into as they are described. This simple psychological need of the reader is sometimes overlooked by the technical writer who has been so absorbed in details that he forgets to write down the main purpose of his design. A list of parts is not a substitute for a definition. The psychologically correct place for it is after, not before, the definition.*

Thus, when you are writing about an investigation or an experiment, start out by stating its general nature and its purpose. When you are describing an apparatus or a machine, tell your reader about its fundamental character and its overall arrangement before you mention any details. When you are setting forth a theory, state its general purport before you start deriving equations.

In short, before you plunge into any discourse, *tell your audience what it's all about.*

Here is an example of the obscurity that results when an exposition begins by describing parts rather than essentials. The left-hand column contains the opening paragraphs of a student report on "The Efficiency of a 24-Inch Cyclone Collector." The right-hand column contains a suggested revision in which the only change is an alteration of the sentence order.

Original Example

A cyclone collector is an apparatus used for the collection of dust particles

Suggested Revision

A cyclone collector is an apparatus used for the collection of dust particles

Original Example	*Suggested Revision*
entrained in gas. The body of the cyclone is a cylinder surmounting an inverted cone. The gas inlet is a rectangular pipe entering tangential to the body of the cyclone; the gas outlet is an inner concentric cylinder extending down into the body. The outlet for deposited dust is at the bottom of the cone. As the gas is drawn through the cyclone, the structure of the cyclone imparts a rapid swirling motion to it. The exact nature of the process by which the cyclone pulls the dust out of the gas is not known, but it is thought to depend on the centrifugal force on the dust particles in the rapidly swirling gas.	entrained in gas. As the gas is drawn through the cyclone, the structure of the cyclone imparts a rapid swirling motion to it. The exact nature of the process by which the cyclone pulls the dust out of the gas is not known, but it is thought to depend on the centrifugal force on the dust particles in the rapidly swirling gas. The body of the cyclone is a cylinder surmounting an inverted cone. The gas inlet is a rectangular pipe entering tangential to the body of the cyclone; the gas outlet is an inner concentric cylinder extending down into the body. The outlet for deposited dust is at the bottom of the cone.

Or again:

Spot-test Analysis. The great value in the use of spot tests lies in the saving in time and materials; they are used where the supply of the material to be tested is small or where quick analysis is vital, as in process control. The amounts of test material used are very small, often only one drop. Spot tests are frequently used in the macro or micro schemes of analysis where identification would normally be poor. A fairly high degree of skill and some special apparatus are necessary in spot-test analysis. Reactions are carried out on filter paper or on spot plates, microcrucibles, centrifuge cones, and micro test tubes. Many effects are used as aids in analysis, such as the capillary properties of filter paper, adsorption, catalyzed reactions, and fluorescence.	*Spot-test Analysis.* Spot tests are individual tests for single elements made on a single drop in the form of a spot on filter paper, a spot plate, a microcrucible, a centrifuge cone, or a micro test tube. Many effects are used as aids in analysis, such as the capillary properties of filter paper, adsorption, catalyzed reactions, and fluorescence. The great value in the use of spot tests lies in the saving in time and materials; they are used where the supply of material to be tested is small or where quick analysis is vital, as in process control. Spot tests are frequently used in the macro or micro schemes of analysis where identification would normally be poor. A fairly high degree of skill and some special apparatus are necessary in spot-test analysis.

A more sophisticated example—and perhaps a more convincing one—comes from a leaflet entitled *United States Government Grants, Manual of*

Procedures and Policies, issued by the Institute of International Education. These are the opening two paragraphs:

> *1. Administration of the Program*
>
> The Secretary of State is responsible to Congress for the administration of Public Law 584, 79th Congress (the Fulbright Act). For the purpose of selecting recipients of awards and of supervising the exchange program, a Board of Foreign Scholarships, as provided by the Act, was appointed by the President. The Board of Foreign Scholarships makes the final selection of all grantees both foreign and American. The Institute of International Education has been requested by the Department of State and by the Board to aid in the operation of the student portion of the program, to publicize announcements concerning opportunities, to receive applications from American graduate students and to assist in the preliminary selection of applicants for student awards. The Institute of International Education also assists in raising dollar support for the tuition and maintenance in the United States of foreign students, recipients of travel grants under the Fulbright Act, and in arranging academic connections for them.
>
> There are three other cooperating agencies concerned with the operation of the program. The Conference Board of Associated Research Councils was designated by the Department of State and the Board of Foreign Scholarships to assist in the preliminary selection of applicants for advanced research or teaching in foreign universities. The United States Office of Education has the responsibility for the preliminary selection of applicants for teaching in national primary and secondary schools abroad. The American Council on Education is responsible for the preliminary selection of applicants for teaching in American primary and secondary schools abroad.

Now, don't you want to know what the program *is* before you are told how it is administered? If you read through to page 6 of the leaflet, you come upon a sentence that begins to tell you:

> The broad purpose of the Fulbright Act is to foster the growth of international understanding by providing opportunity for representative Americans to live and study abroad for an academic year.

Contrast this order of presentation with another contained in the same publication. On page 15 is a "Sample Press Release":

> Opportunities for more than 600 Americans to undertake graduate study or research abroad . . . under the terms of the Fulbright Act have been announced by the Department of State. Countries in which study grants are available are Australia, Austria, Belgium, Burma, Egypt, France, Greece, India, Iran, Italy, the Netherlands, New Zea-

land, Norway, the Philippines, Thailand, Turkey, and the United Kingdom.

The awards will enable students in all fields of graduate work and those with specialized research projects to study in foreign institutions and universities under renowned professors and specialists. Grants also are available to students with records of accomplishment in such fields as music, art, architecture, and drama. A few opportunities in workers' education and social work are provided in the United Kingdom.

The grants are made under Public Law 584, 79th Congress, the Fulbright Act, which authorizes the Department of State to use certain foreign currencies and credits acquired through the sale of surplus property abroad for programs of educational exchange with other nations. Grants are normally made for one academic year and generally include round trip transportation, tuition or a stipend, a living allowance and a small amount for necessary books and equipment. All grants under the Act are made in foreign currencies.

Don't you get a clearer picture and an easier one to assimilate when the broad, basic definition comes first? After this orientation, you are ready to be told about the administration of the program.

Here is another example in which the general nature and purpose are stated at the beginning, this one from the organizational manual of a telephone company.

FUNCTION OF THE ENGINEER OF BUILDINGS

General

The general function of the Engineer of Buildings is to plan for the building requirements of the Upstate Area for central office equipment, test bureaus, business offices, and quarters for personnel; office buildings; and company-owned garages. Included are the selection of adequate land, development of study plans, and preparation of architectural plans and specifications, including all mechanical and electrical work. The Engineer of Buildings is also responsible for providing consulting and advisory service to the plant forces in connection with the maintenance of buildings and building equipment. Serving under the Engineer of Buildings are the following:

1. Planning Engineer

(1) Determines the size of lots for land purchases. Makes studies of lot usage and building space layout for initial and ultimate equipment and personnel requirements, obtains recommendations for operating departments, and submits preliminary study plans for approval of departments.

(2) Prepares final study plans for use of the Building Engineer and the Chief Architect and checks plans.

2. Building Engineer

(1) Inspects land to be purchased for new buildings and reviews study plans for architectural and engineering adequacy. Prepares architectural and structural plans for buildings; engineers all types of mechanical equipment in buildings, such as power service, lighting, heating, plumbing, and ventilation; and supervises the construction work in the field. Employs the services of architectural firms for the larger building projects.

(2) Prepares and maintains a budget of proposed building expenditures, prepares specific requests (estimates) for authorization to cover the cost of building projects, submits construction work for bid, approves bills, and clears them to the Accounting Department for payment. Maintains an inventory record of all building property and a record of drawings, photographs, and specifications.

(3) Acts in a consulting and advisory capacity with the operating departments regarding building maintenance; reviews and approves structural changes.

(4) Reviews systems-initiated engineering letters and practices, determines standards, prepares practices as required for area use, and arranges for appropriate distribution to the field. Reviews insurance inspection reports and forwards them to the field, advising the insurance inspectors regarding disposition of recommendations.

In this example, consistency of headings and parallelism help to maintain a logical organization of the material.

Remember: before you plunge into any discourse, *tell your audience what it's all about.*

2.5 Emphasis of the Significant

Closely related to the ideas of Sec. 2.4 is a fault found in technical reports perhaps more commonly than any other: the burial of the meat—the really important and significant ideas—under a mass of details. The reason for this failing is inherent in the nature of technical work. Before he finishes a task, the technical worker is thoroughly familiar with the main line of thought—the purpose and the method of attack. From day to day he has been coping with a lot of smaller problems, many of them probably routine and tedious. Consequently, when he sets about reporting on his work or his investigation, he talks primarily about the little problems that have occupied so much of his time and attention.

It is scarcely necessary to tell the engineer and the scientist how important it is to get the main points out in the open so that they will be seen and recognized and acted on. But it does seem in order to warn him that *he must usually make a conscious, planned effort to keep his key ideas uncovered.*

The remainder of Sec. 2.5 suggests some specific devices for making the important material stand out. In addition, practically all of the principles propounded in the rest of this book will help you to accomplish the same vital end.

2.5.1 Prominent Position

Perhaps the most fundamental way of making an idea stand out is to put it in a prominent position. The most prominent position in any report or paper is the *very beginning*. Even though you have not yet led up to it logically, you can often put across your major thesis most effectively by stating it right at the start and later supplying support for it.

A secondary prominent position is the end, particularly in a short communication. In a longer report or paper—we may as well face it—your reader may never get to the end.

If you want to bury information or an unpopular opinion, put it in the middle part of a report. The chances are that it will not be read.

Some specific suggestions for emphasis by position are given in Chapters 6 and 7, which discuss the structure of reports.

2.5.2 Elimination of Detail

One sure way to stress important information is to *remove* unessential material. Perhaps you can scrap it, a difficult thing to do to the product of your toil.

On the other hand, you may need to include a lot of details for record purposes. If you do, you can usually put them into an appendix, leaving your main discourse uncluttered.

Ways of eliminating material are discussed more fully in Sec. 2.6; the function of the appendix is described in Sec. 7.4.

2.5.3 Elimination of Words

The clearing away of superfluous words almost always allows more light to reach what is left; see Sec. 14.3.

2.5.4 Liberal Use of Headings

The great value of headings, or heads, as they are often called, is discussed in detail in Sec. 2.8. One of their important functions is to announce and call attention to key material.

2.5.5 Repetition

Psychologists tell us that children learn by repeated experience. The same process works on even the sophisticated and highly educated scientist or engineer. If you want to make sure that an idea does not fade into the background, repeat it. Programmed learning, one of the *newer* communication techniques, uses repetition as a principal technique.

Redundancy is a well-known engineering term. Repetition is a form of redundancy, the deliberate repeating of information to emphasize an important point. Repetition is used in the introduction, the discussion, and the summary of reports, and in the statement of purpose and the technical description of proposals; and it is a most useful device for getting points across in the body of the article. But repetition must be the result of a definite design on the part of the author, not of carelessness or forgetfulness.

If you can say something in a different way each time, the repetition will be subtle and palatable; but don't shy away from even bald, frank repetition if your point is important enough to warrant it.

2.5.6 Visual Aids and Tables

Visual aids—graphs, curves, drawings, diagrams, photographs—often present information in a striking and efficient manner. They reinforce and emphasize key ideas.

The various visual aids, and tables, can themselves suffer the fault of hiding the significant behind a maze of unimportant detail. Ways of making them bold and clear are described in Chapters 18 and 19.

2.5.7 Typography

Capital letters, larger type size, boldface, and italics are all effective means of emphasis *if not overused.* Of these, only capital letters and italics (underlining) are available on the typewriter.

2.5.8 Specific Mention

It is all right to be explicit and say, "This is a particularly important point," or "This is an important part of the report." Such statements may be made along with the material they refer to, or they may be made separately in a letter of transmittal or preface.

2.6 How to Be Brief

Brevity may be achieved in two different ways: one is to skim over all the material at hand; the other is to cull out the insignificant and treat the significant exhaustively. This culling process is not to be confused with suppression of data in order to color conclusions, a device that is not practiced by any self-respecting engineer or scientist.

Consider the two bar charts of Fig. 1, page 14.

Let us say that each bar represents the treatment of one factor in a paper you have written. A 100-percent complete coverage means that you have supplied so much information that the reader can understand you with a minimum of work on his part—without having to fill in any gaps or steps.

Now suppose that your first draft of the paper is obviously much too long —let us say twice too long. You must remove half of your draft.

You can conceivably remove steps and particulars from the treatment of each of your five factors until they are each just half as long as originally—and each just 50-percent complete. The result is depicted in Fig. 1a.

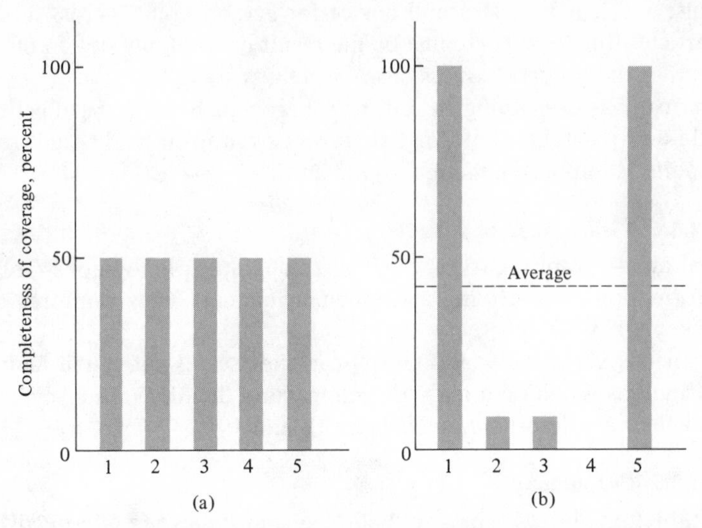

Fig. 1. Ways of achieving brevity: (a) skimming; (b) culling

What effect does this boiling-down have on your reader? Each of your five factors is described fully enough to make him try to understand it. Yet he must go to endless work to fill in the information you have withheld from him. If he is very curious about your subject, he will go through with this tedious process, with a large expenditure of effort and great annoyance. But more likely he will give the whole thing up as a bad job and turn to some easier and pleasanter task.

On the other hand, suppose you appraise your five factors carefully. You decide that factor 4 is trivial and can be thrown out without serious loss. Factors 2 and 3 should be very briefly mentioned. And factors 1 and 5 are really significant, deserving of 100-percent coverage. Again, as shown in Fig. 1b, the average coverage, and thus the total length, is about half the original.

And now what of your reader? Your brief mention of factors 2 and 3 does not stimulate him to fill in some gaps and find out all about them; he is willing to accept them on your say-so. Factor 4 is, of course, no worry to him at all. And the significant factors 1 and 5 he understands fully and with a minimum of work on his part.

Only two levels of treatment are readily accepted by most people: the very complete and the very brief. As illustrated in the following short example, the great middle ground of partial treatment tickles the curiosity, then annoys or baffles.

> Many metals are extracted from their ores by a process known as smelting. For example, iron ore is placed in blast furnaces with coke and limestone. The ore is heated, reducing the metal to the free and molten state. The molten iron is poured into molds.

The reader unfamiliar with the smelting of iron ore is left wondering how the coke and limestone enter into this process. The writer should have either explained why they are put into the furnaces with the ore, or not mentioned them at all.

Since most of the readers of this book will be college students, we should like to call attention to a serious departure of the usual college *theme, term paper,* or *laboratory report* from the principles of good technical reporting we have discussed up to this point. The primary purpose of these college papers is usually not to transmit some information to a specific audience; it is, rather, to show the instructor that the writer has gained some knowledge. The paper is addressed to somebody who already has the knowledge, and who is thus by no means a typical audience. To make sure that they omit no pertinent information, the writers of college themes and papers tend to include every detail they can lay their hands on, rather than to cull out the unimportant and the insignificant.

Consequently most college papers are not good exercises in technical communication. The provision of realistic reporting assignments for technical students usually requires a carefully planned program of cooperation between the English department and the several professional departments.

2.7 Organization

Unfortunately, there is no neat formula for the organization of technical reports. Each report must be organized to fit its own subject, its own purpose, its own audience. But a few general principles apply to most technical communications.

2.7.1 Logical Progression toward Conclusions

Almost every technical report aims to bring its reader to certain conclusions. These conclusions need not be of the conventional, formal variety such as "It is concluded that the piston failed because of an inclusion in the casting." They may be simply that "No conclusions are warranted by the data developed in this study." But in general, the information you present in technical reports is intended to bring your reader to some sort of a conclusion.

Thus the material in any report or paper should be presented in an order that leads logically toward a conclusion or conclusions. This does not mean, of course, that everything in a lengthy report will aim at one final climax; the various sections of the report are organized so that each of them has its logical conclusions.

Application of this rule might seem to relegate the conclusions—usually the most important ideas in a report—to a position of relative obscurity, buried under pages of data. To make sure that the conclusions are not overlooked by even the busiest or most hurried reader, most present-day technical reports and papers carry a statement (or a restatement) of the important conclusions at the very beginning. See Secs. 6.1, 7.2.4, and 7.3.

2.7.2 Topical vs. Chronological

The two basic approaches for any exposition are the *chronological* and the *topical.* Except for the purely historical subject—and sometimes even for it—the topical treatment is preferable in technical reporting. By and large, your reader wants to know what you investigated and what you found out—he wants to know why you did things and how you did them—rather than the story of your investigation, or a *narrative,* as it is sometimes called in industry.

Of course some chronology may be entirely appropriate. The events leading up to an investigation may show why you have undertaken it, or a procedure may appear more sensible and more logical if you describe the order of the steps you went through.

Although you may have occasion to mix some chronology into your treatment, the topical approach will usually be basically the more suitable one.

2.7.3 The Three Parts

Almost every technical communication should have three functional elements. This does not mean that it should be divided by boundaries into three distinct parts. But functionally it should have a beginning, a middle, and an end. Too many technical reporters devote so much of their attention to the middle that they slight the beginning and the end, both of which are very important.

The beginning is often called the *Introduction,* although a more specific title is preferable (see Sec. 7.2.2). This beginning orients the reader. It supplies him with background material, so that he will see how the subject of the paper fits into the general scheme of things. It helps to supply the framework called for in Sec. 2.4. It prepares the reader for the main presentation of information—the middle.

The middle, often called the *Body,* is usually the longest part of the report. It can be organized in many different ways; some of them are discussed in Sec. 7.2.

The end is sometimes labeled *Conclusions,* although again a more specific title is preferable. It brings together the various subjects that have been discussed and shows their relationships with each other and with broader fields. It leaves the reader with some thoughts about the main subject under discussion, rather than merely about one phase of it. This end section makes the exposition come to a logical and an obvious termination, rather than simply stop on a note of detail. It ties a string around the bundle.

2.8 Headings

The importance of headings and subheadings can hardly be overestimated. You will seldom see too many of them in a technical report or paper; very often, too few. They serve a number of vital functions.

2.8.1 A Help to the Reader

(1) Subheads make the structure of the exposition apparent. They help to supply the reader with the framework on which to fasten the parts.

(2) Subheads serve as convenient and efficient signposts. They let the reader know that he has reached the end of one subject and is about to begin a new one.

Two alternatives occur if this information is not supplied by subheads: either the writer supplies it in sentence form, as part of his exposition; or the reader is left to discover belatedly that he has left one subject and embarked upon a new one. The first of these alternatives wastes the reader's time; the second wastes his time and temporarily mystifies him besides.

(3) Subheads permit the reader to interrupt his reading and yet pick up the thread of the discourse when he returns to it. With a piece of writing that has a generous supply of subheads, the reader has to go back only a few paragraphs to recapture his thoughts. But when he returns to an exposition with few subheads, or none, he may have to backtrack for several pages before he can proceed again.

(4) Subheads help the reader also when he must reread long or complicated expositions in order to refresh his memory or to cast new light on a subject only partially understood at first. He can find the material he is looking for very much more easily if he is supplied with subheads; moreover, they enable him to skim and skip.

The mind, like the stomach, can assimilate material more easily if it is broken down into chunks of digestible size before it is swallowed. Subheads perform this chewing function very efficiently.

2.8.2 A Help to the Writer

(1) The logical system of subheads, by acting as an outline, keeps the writer in clear channels. (The addition of subheads to an exposition that did not originally have them often brings to light repetition or other illogical organization.)

(2) The use of subheads is a particularly easy way of supplying the transitions, or connective tissue, needed to make any discourse smooth and unified.

(3) Subheads make it easier to refer to specific parts of a report or paper.

Obviously, the longer and more complex any exposition, the more it needs subheads. But even the very short report can often be clarified and made easier for the reader if it is broken up by subheads. A half-page, two-paragraph memorandum reproduced in Sec. 21.2.2 is certainly helped by the presence of two subheads.

2.9 Consistency

Technical writing should be thoroughly consistent in every phase. If you label a point Q on a drawing but call it q in the text, you will confuse and mislead your reader. If you refer to a part at one moment as a pivot and at the next as a pin, he may think you are talking about two different objects.

For example, a recent paper on the design of an electrical measuring instrument discussed the types of circuits that might be suitable. In various parts of the discussion the following types were mentioned: (1) condenser-resistor, (2) resistor-inductance, (3) capacitance-resistance, (4) resistance-condenser, (5) inductance-resistance. The thoughtful reader can reduce these to two kinds of circuits and can pick a consistent pair of terms for them—*capacitance-resistance* and *inductance-resistance*. But he should not be put to this trouble. (As a matter of fact, the paper from which this example was taken used still another term—the ambiguous *reactance-resistance*—to mean inductance-resistance.)

A particularly troublesome inconsistency is a change in units of measurement. For instance, in the two reports on a pair of related tests performed at a commercial laboratory, a wear rate was given first in thousandths of an inch per 1000 miles, then in thousandths of an inch per 10,000 miles. Such a discrepancy can cause serious misunderstanding on the part of a casual reader. Similarly, curves that are to be compared should be plotted to the same scales.

Perhaps less serious than inconsistencies in nomenclature or units but still troublesome and annoying to the reader are inconsistencies in abbreviation, hyphenation, capitalization, or use of numerals. At the very least, the reader pauses to question the variation that has no significance. You owe it to him to reread your manuscript very carefully to see that you have been consistent in even the smallest mechanical details. Sometimes even a consistent small error is preferable to vacillation between a correct form and an incorrect form.

2.10 Specific Identification

You can save your reader a lot of trouble by making every reference and every statement just as specific as possible. On the next page are some non-specific statements with their specific counterparts. Notice how much more helpful the entries in the *Specific* column would be to any reader.

Nonspecific	*Specific*
As described elsewhere in the report . . .	As described on page 6 . . .
These details are shown in the figure.	These details are shown in the center portion of Fig. 8.
Fuel consumption was 15 percent less in Test 12–4Y than in Test 12–3Z.	The Rumbler 8 consumed 15 percent less fuel than the Wonder 6.
The first of these methods is the simplest.	Cupellation is the simplest of these methods.
Fig. 2 summarizes the picture with respect to dealers.	Fig. 2 shows the distribution of retail volume, wholesale volume, and number of outlets among dealerships of various sizes.
If the transaction with Mr. Smith cannot be effected . . .	If Mr. Smith does not buy the bulldozer . . .

2.11 Expression of Opinion

In an effort to stress the importance of accuracy and objectivity in technical reporting, some instructors convey the idea to their students that the technical report must present only facts, never opinion. However, engineers and scientists are employed not only to discover facts, but also to make deductions from those facts and to make decisions based on them. Therefore the technical report must often present the judgments and opinions of its writer or his organization in a *discussion* section. Supervising editors of reports tell us that this section is the most difficult to organize and to present logically.

Your duty is not to refrain from expressing opinion; it is rather to tell your reader very certainly what you are doing whenever you have occasion to introduce opinion into your reporting. Of course your opinion should be based as soundly as possible on demonstrated facts, and it should be as impartial as you can make it. But no matter how objective you think you have been, whenever you express an opinion, tell your reader. Some ways of fulfilling this obligation are discussed in Sec. 14.13.

2.12 Conclusions and Recommendations

Some young engineers are so unsure of themselves that they hesitate to draw conclusions or make recommendations on the basis of their observations. Indeed, in some organizations only the supervisors or executives are permitted to formulate conclusions and recommendations. But unless you have been told specifically that you are to report only observed facts and that you are not to make deductions from these facts, your comments and conclusions will probably be welcomed by your superiors. But be sure never to draw conclusions on the basis of numerical differences that are not significant or that are smaller than the probable error of measurement. Do not be afraid to make the conclusion "No conclusions are warranted by the results of this investigation."

2.13 Date

The date is a seemingly insignificant item that should go on every written record or report. No matter whether it is a highly polished formal report, a memorandum, or a page in a laboratory notebook, it should be dated. No matter what firm you work for or what forms it uses, be sure to date everything you write. For it is often important to know which of two reports was written first, or when a certain practice was started or stopped, or when an action was taken. And of course the date is essential on any record of research or design work that may later become the basis for a patent application. Many a technical man has regretted afterward that he didn't date a memorandum that seemed trivial when it was being written. Remember: put the date on everything you write.

2.14 Titles

In the course of reporting on technical work, you will often have to write titles—titles not only of papers or reports, but also of drawings and diagrams, graphs and curve sheets, tables, and columns in the tables. Every title —no matter which of these categories it belongs to—should be just as short as possible yet still descriptive of its subject. The title of a report or paper should bring out the fundamental, overall nature of the subject, and it should be as specific as possible without going into details. But sometimes the subject is so complex, or so limited, that it cannot be adequately described in a short title. In this event, a title plus subtitle is a useful combination: the brief main title serves as a convenient handle, while the subtitle furnishes specific limiting, qualifying, or explanatory information.

Titles in technical reporting seem to deviate from these ideals in both directions: an occasional title is so brief or so general that it does not sufficiently describe or limit its subject; a good many titles are so detailed that they obscure basic ideas. Here are some examples of faulty titles, with suggestions for their improvement:

Original Example	*Suggested Revision*
Aluminum Oxide	Aluminum Oxide: Physical Properties
Input Devices	Conversion of Shaft Position to Binary Number Code for a Digital Computer
Rumbler V-8 Exhaust Valve Performance During Cycling Operation of a 400-Cubic-Inch Engine with Fuel Containing 5.0 ml Anti-K per Gallon and with S-L Synthetic Crankcase Lubricant	The Effect of S-L Synthetic Crankcase Lubricant on a Rumbler V-8 Engine

Many reports carry titles encumbered by superfluous words that fall into a pattern: *Report on . . .* , *Study of . . .* , *Test of . . .* , *Investigation of . . .* , and so on. The information contained in these phrases is usually self-evident. If a report is put up and bound in the conventional way, it is quite obviously a report, and nobody is enlightened by the words *Report on* on its label. Similarly, does *Investigation of the Physical Properties of Aluminum Oxide* convey any information not contained in *Aluminum Oxide: Physical Properties?* Nobody is likely to report on these properties without first investigating them. Usually words of this pattern can simply be deleted from any title; occasionally they may be significant enough to be left in.

2.15 Test for an Effective Report

The foregoing sections of this chapter have suggested a number of basic principles for effective technical reporting. Most of these principles apply equally to written or oral communications, long or short ones, formal reports, informal reports, papers for the journals, or articles for the press.

We have now established enough of the criteria of good technical reporting to suggest a quick test for the effectiveness of any technical reporting job. When you have finished reading it (or hearing it), ask yourself *what is its central theme*. If you can answer this question readily, the report is probably a good one.

EXERCISES

1. The following excerpt represents one part of a monthly report. Reorganize it to illustrate the ideas expressed in Sec. 2.5 and rewrite it to make it clear and concise.

SMOKING PROJECT—ANDERS HIGH SCHOOL

> Anders High School was the scene on April 24 of a meeting between the school officials and the health-education consultant. Social and cultural aspects of smoking were discussed to reduce peer group pressure to smoke and to expose smoking for what it is, an unsafe habit.
>
> In order to do this, it was suggested that PTA and parent groups should hold seminars and have them work with school personnel in reducing teenage smoking. It is necessary that the child get an elementary understanding of the probability of statistics—that he is more likely to die of lung cancer if he smokes than if he doesn't.
>
> Work should be done with reference groups of youth, those to which he belongs, since a person's attitudes of behavior are strongly influenced by the groups that offer reference perspectives to him.
>
> Parents are extremely important people to a great majority of teen-

agers. Parents are the role model for their children. There is a relationship between parents' smoking and a child's smoking.

The above recommendations should be implemented as soon as possible.

2. In order to use the information in later exercises and in oral presentations, prepare a questionnaire to determine the technical and scientific level of your class. Write a summary of the results.

3. The beginning of most reports is called the *Introduction*. This is a vague term that seems to mean different things to different people. Select three technical reports. Outline the introductions and compare them. Write a summary of the results.

4. Go to the library and obtain an industrial or research report. Make a check list of the important fundamental principles for preparing a report. Place them in a left-hand column. In the right-hand column, evaluate the report according to these principles.

5. Write a short account of two reports you have looked over. One of the reports should be an example of the chronological approach, the other of the topical approach.

6. Find examples in existing reports of how the authors have handled the basic principle of stating the general nature and purpose of an investigation or an experiment. Report orally to the class on your study.

3. General Procedure

People often ask how to go about writing a report or paper. Again, unfortunately, no formula is available: each person has his own work habits, and each subject imposes its own requirements. But a general order of procedure does seem worth suggesting.

3.1 Collection of Information or Data

The first requisite for good writing or good speaking is to have something to say. So the first step in every technical reporting job is the collection of data or the development of information.

In general, information comes from these sources: *documentation, experimentation, observation,* and *logical thinking*. It may be gathered in a laboratory or a library; it may come about through the observation of work being done; or it may be the product of logical thought processes in the mind of the writer.

When you are writing, you must learn to judge and to evaluate your material. Certainly one of the first principles is to approach your material in a scientific manner. Because your emphasis will be on the material itself and not on any preconceived ideas, be ready to shift your point of view, and the scope of your writing, up to the time you put everything down in a first draft.

No matter what the source or nature of your facts, you can facilitate one of your early steps—the preparation of the outline—by recording the facts on cards, one topic to a card. Take complete and accurate notes as you go along, whether they come from the laboratory or the library. The usual practice is to put the data on cards of reasonable size (4 x 6 inches or 3 x 5 inches is handy). But remember that the most important single item in note-taking is accuracy. Verify quotations, and especially figures, while the sources are available.

One way of recording data is to break your topics into reasonable subtopics and to list each of these on a card, starting at the top on the left side. The main portion of the card will then contain the information itself. How complete it is depends on how readily available your sources are. For example, if you cannot retain a source book or photocopy of it, or if you cannot trust your memory beyond a certain time, your notations will have to be particularly complete; otherwise, scattered phrases may be enough. At the bottom of the card, cite the source, with page numbers if the note came from a document.

Here is what a note card might look like:

> Applications of Permanent Magnets
>
> Permanent magnets can be classified in several ways: mechanical-to-mechanical attraction and repulsion, mechanical-to-heat, mechanical-to-electrical, and electrical-to-mechanical.
>
> B.R. Neris, "Permanent Magnet Assemblies." Mechanical Design (January 18, 1967), 128-132.

A word of caution about library research: When you are gathering information from the library, you should probably record it on your cards in outline or telegraphic style. This method will preclude the inadvertent plagiarism that might occur when you have later forgotten whether you copied or paraphrased full sentences.

You may have occasion to copy some passages verbatim, because you may want to quote an author's own words. Be sure to put prominent quotation marks on any cards that carry verbatim quotations.

Note cards, of course, will deal with other information besides that found in books. Speeches you have heard, material from interviews, digests of conferences, trips you have taken through factories—all of these can be reduced to a workable form by recording them on cards.

3.2 The Outline

If you use this system of topical note-taking, you are well on the way to making an outline. Because outlines result from a combination of what you already know about a subject and what you record on your note cards, very early in your writing you should draw up some kind of tentative outline.

If you are human, you will probably depart from the outline at one place or another when you come to do the writing. Setting ideas down on paper usually requires rigorous thinking that further clarifies them in your mind or sheds new light on them. You may have to revise your outline as you write.

But the outline is nonetheless a valuable tool. At the very least, it provides you with a starting point, serves as a general framework, and keeps you thinking along logical lines.

The longer the paper, the more you will need an outline; but even a short composition will be easier to write and clearer to read if you construct an outline before you set down a single sentence.

The entire process of outlining, then, usually takes place in these four stages:

(1) A tentative outline is set up at the earliest possible time.

(2) Gathering of material continues and a topical card system is built up.

(3) Note cards are arranged and rearranged in the light of the tentative outline and the data compiled.

(4) A final outline is made and becomes the basis of the rough draft.

The function of the outline does not end with the writing of the text. The reader of a really well-planned report or paper should be able to detect the outline, at least its major divisions. Perhaps he can reconstruct it simply from the flow of topics; perhaps you will supply it for him by using your outline entries as subheadings in the final draft.

Some suggestions for arrangement of outline headings are given in Sec. 17.7.

3.3 Tabulation and Plotting of Data

Whenever you are writing about any subject that involves numerical data, you should probably tabulate or plot your figures before you start writing. This sequence of operations will help you to interpret your facts and to write about them clearly (see Chapters 18 and 19).

3.4 The Rough Draft

After you have constructed your outline, *write a rough draft just as quickly and just as consecutively as you can,* paying little attention to niceties of language. Leave plenty of space between lines for subsequent revision. *Concentrate on a smooth and connected presentation.* You may have to go back momentarily and reread sections or even revise them slightly to keep your approach consistent, but save the major polishing job for later. This sequence will help you produce spontaneous, integrated, and coherent papers and reports.

3.5 Revision and Polishing

Have you ever read a letter or a paper that you yourself wrote a year or two previously? Did you notice how objectively you were able to view it, as if you were reading the writing of someone else? Perhaps you were pleased with the general effect of your brain child; but very likely you came upon passages that were not entirely clear because of gaps in the information supplied, or you were struck with awkward repetition of words, or you noticed other matters that you would change if you had the job to do over again.

When you read a passage right after you have written it, your complete familiarity with the subject enables you to fill in the gaps that you have left in your exposition. But in time these pieces of the puzzle escape from your mind, and then you, as a reader, are confronted with the same problems that a stranger would be.

Therefore you should *let as much time as possible elapse between the composition of the rough draft and the revision-and-polishing job.* (We hasten to point out that this procedure entails writing the rough draft earlier, rather than the final draft later.)

Professor Douglas H. Washburn of Rensselaer Polytechnic Institute has compiled a check list to be used in revising a manuscript.

In part, he offers these suggestions for revising, polishing, and evaluating your rough draft:

(1) State in a single sentence what you wish the reader to retain. Write this sentence out. It is intended to define the central communication you wish to achieve and to which all the details are subordinate.

(2) Read the draft straight through in order to decide in what respects it needs to be revised.

(3) Make up your mind in which of the following ways you wish to revise:

(a) Shorten the manuscript.
(b) Increase the readability.
(c) Make the organization more effective.
(d) Make a closer tie-in with illustration.

(a) To shorten the manuscript, look for:

Circumlocutions: "*In the month of May* production reached a peak." (See Sec. 14.3.2.)
Superfluous words: "The engine *which has been equipped* with a governor . . ." (See Sec. 14.3.1.)
Phrases to be reduced to single adjectives or adverbs: "The sample *with the coat of red* paint . . ."
Adjectives or adverbs the opposite of which would be unlikely in the context: "One *possible* plan would be to . . ."
Ideas or facts that do not contribute to the main part of your writing, as defined in the single sentence.
Examples demonstrating points already made clear.

(b) To increase readability:

Break up long sentences. (See Sec. 14.5.2.)
Find common words to substitute for your more unusual terms. (See Sec. 14.2.2.)
Use more headings. (See Sec. 2.8.)
Place your main points in conspicuous positions. (See Sec. 2.5.1.)
Show your reader how the subject being discussed affects him.

(c) To make your organization more effective:

Check your rough draft against your outline.
Make certain that your table of contents is logically arranged.
Check your paragraphs to see that they are logically constructed and deal with one topic at a time.

(d) To make a closer tie-in with illustrations:

Make sure that each illustration is placed immediately after the first reference to it in the text. (See Sec. 19.3.)
Replace such stock phrases as "See Fig. 96" with a discussion of the illustration to guide the reader to a complete understanding of the text-illustration relationship.

The amount of polishing required depends on the writer's ability to say things well the first time and also on the use to which the composition will be put. A newspaper article or a memorandum cannot be delayed for the overhauling that you would accord an article for one of the technical journals, nor is it likely to be worthy of such effort.

Many experienced writers go slowly and painstakingly through several drafts before they are satisfied to release any serious piece of writing for publication. Dull as the task may sometimes seem, you should by all means do a thorough job of revising and polishing everything you write for more than casual or transitory purposes.

3.6 Typing

After final revision, your report or paper will be typed. You may have to do this task yourself; you may be fortunate enough to have a wife who will do it for you; or you may be in an organization that supplies stenographic service. The typed copy may be the end product of your labor, or it may be the basis for reproduction of your opus in quantity. (Methods of reproduction are described in Sec. 17.11.) In either event it should be as accurate and as neat as possible.

3.7 Proofreading

If you do your own typing, you will probably make typographical errors. If you have your typing done by someone else, even a professional stenographer, remember that she has probably copied much of your manuscript without understanding what she was typing, particularly if your treatment is highly technical or mathematical. This kind of mechanical copying is likely to lead to errors even by a highly skilled typist.

Therefore, no matter who has typed your final copy, it will probably contain mistakes. The only way to eliminate all of them is for you yourself to do a thorough job of proofreading.

You owe this admittedly tedious chore to yourself and to your readers. Remember: it is not only the major blunder that is confusing or misleading; the minor typographical error can distort or even contradict your meaning. Your reader should never be put to the trouble of straightening out even a fairly obvious slip; and you should preserve your reputation for accuracy and precision by seeing that he never has to.

4. Technical Description

Frequently in technical reporting you must describe not only the complete project but also machines and mechanisms with which you worked, processes within a larger framework, and theories involving both machines and processes. An engineering writer not only finds technical description an indispensable part of the report, but he also uses it in manuals and proposals, and carries it over in more sophisticated form into papers and articles.

4.1 Description of Machines and Mechanisms

A machine is a working object; consequently you must show the reader not only what it is, but also what it does. Give him a general statement incorporating these two points to place the device in functional surroundings.

You may have to describe the theory or principle upon which the machine is built. In general, you can assume that the specialist will be familiar with the principles of machines that are modifications of those already in use; that the background principles of radically new machines should be extensively explained; and that for the more general reader a discussion of theory should be omitted.

Unless you are describing an extremely simple machine, do not take up details before you have given a broad, overall description (see Sec. 2.4). It is easy for the reader to get misconceptions about size, shape, color, and other physical characteristics.

Details should be described in a systematic manner according to some logical order. One possible order is by *function;* that is, the parts of a machine are described in the order in which they function when the machine is put in motion.

Sometimes an order of *importance* is more logical. Although a machine usually depends on all its parts, some of these may be more important than others because they represent new ideas put into practice, or because certain primary operations depend on them. Occasionally you will be faced with the fact that two or more parts operate at the same time. For any of these reasons, you may have to fall back on *importance* as a unifying principle.

A third logical order often used is *spatial.* The parts are described as you come to them. This may be from left to right, top to bottom, inside to outside.

Here is an example: a description of a mechanism used on some automobiles.

THE QUIK-STOP ANTI-SKID BRAKE SYSTEM

The Quik-Stop anti-skid automobile brake system is a wheel deceleration-sensing and compensation system that prevents wheel lock-up by alternately applying and releasing the brakes. Experiments show that the highest retarding friction between tire and road is obtained when slip is about 15 percent. The Quik-Stop system maintains slip within close limits on each side of this value.

The system has three basic elements: (1) the sensors, which monitor the velocity and deceleration of the wheels; (2) the electronic module, which processes this information into a usable signal; and (3) the actuator, which modulates the hydraulic pressure supplied to the brakes.

The sensor consists of a toothed ring rotating with the wheel, a static toothed ring, magnets, and a coil. When the teeth of the two rings are in coincidence, flux builds up; when the teeth are out of coincidence, the flux decays. The changing flux induces a sinusoidal voltage output from the coil whose frequency is proportional to wheel speed.

The electronic module receives this signal and counts the pulses much as a frequency meter does. The signal is amplified in the module and sent to the actuator.

When an appropriate signal reaches the actuator, a solenoid opens an air valve, admitting atmosphere to one side of a vacuum-and-spring-suspended motor and causing the diaphragm to compress the spring. A piston in the hydraulic system retracts, closing a check valve and isolating the wheel cylinder from the master cylinder. Further travel of the piston withdraws fluid from the wheel cylinder, reducing pressure. When a re-apply command is received, the solenoid closes off the atmosphere and reopens the vacuum source, causing the diaphragm and piston to return to normal. That re-establishes connection between master cylinder and wheel cylinder, so that hydraulic pressure is again applied to the wheel cylinder.

The whole cycle is repeated at from 6 to 8 times per second, effectively preventing any protracted wheel lock-up and maintaining braking forces at a maximum no matter what the friction condition of the pavement.

4.2 Description of Processes

It is difficult to imagine a process that does not involve a machine or device of some kind. Therefore, what has already been said about describing machines applies to processes. The major difference, however, between describing a machine and a process is that with a machine the emphasis is on the object that does the work, whereas with a process the emphasis is on the work itself. You must, therefore, take special care to see that the steps in a process are carefully explained and arranged. Your material will determine your method. As with machines, you can use an arrangement based on *function, importance,* or *space,* or combinations of these.

Sometimes there are several ways of carrying out the same process, especially with natural phenomena. Soil erosion will produce the same result, but the process in one locality may be quite different from that in another. In your description, then, determine whether there are variations to the basic process. If so, they should be described. And it will clarify your writing if you end with a complete description of the product obtained or any other end result of the process.

The following procedure is part of a larger process. It should give you some idea of the steps involved in describing how a certain result is arrived at.

PREPARATION OF METAL PARTS FOR
DRY-FILM LUBRICATION

Dry-film lubrication is an efficient method for lubricating metal parts. Dry-film lubricants are mixtures of lead, tin, and either graphite or molybdenum disulfide suspended in a thermosetting resin binder and sprayed on the parts to be lubricated. The binder is usually a vinyl chloride-vinyl acetate resin combined with a thermosetting phenolic.

At present two types of dry-film lubricants are being used: Electrofilm No. 6281 and Electrofilm No. 3849. The final use of the part to be sprayed determines the type of lubricant to be applied.

Before metal parts can be dry-film lubricated, they must be subjected to a surface preparation involving a cleaner and several baths.

The following procedure is used:

1. Cleaning

The metal parts to be worked are immersed in the emulsion cleaner at room temperature. The *cleaner* is an emulsifiable solvent, Enthose 75. It is used as furnished and is not mixed with water. Light oils are immediately penetrated and work can be withdrawn in a few seconds. Heavier oil films and dirt contamination require longer soaking time.

After the parts are drained, they are thoroughly rinsed with water and are then ready for a pickle bath.

2. Pickle Bath

The *pickle bath* is used to dissolve rust scale rapidly and efficiently without attacking the exposed steel surfaces. This bath consists of Okite Compound 32, cut 50% with water.

After being put in the pickle bath, the parts are rinsed before immersion in the pre-phosphate bath.

3. Pre-Phosphate Bath

Any contamination remaining after the completion of the cleaning and pickle cycle is removed in the *pre-phosphate bath.* The pre-phosphate bath is prepared by adding 100 cc of cp concentrated nitric acid to 100 gallons of 85% phosphoric acid. The immersion will decrease the time in the phosphate bath.

The parts are again rinsed thoroughly and are ready for the next step.

4. Parcolene Z Bath

Parcolene Z chemical is used as a conditioner to promote the formation of a dense and finely crystalline phosphate coating. The conditioning solution is prepared by adding 8.5 pounds of the chemical to 100 gallons of water.

In this step, the parts are treated with Parcolene Z chemical for 10–60 seconds at room temperature. The bath must be agitated to prevent settling of active ingredients.

After the conditioning, the parts go directly into the phosphate solution without a water rinse.

5. Phosphate Bath

The parts go into a *phosphate bath* to create a complex zinc phosphate coating. The bath is a 2½% solution of Cripcot HC in water, operated at a temperature of 180–200 F.

After the bath, the metal parts are given a hot-water rinse and dried immediately. They should be sprayed with dry-film lubricant as soon as possible to prevent absorption of moisture from the atmosphere.

4.3 Description of Theories

Theories are usually more difficult to describe than machines or processes. With machines and processes, the material is based largely on facts, and your primary purpose in describing them is to inform. But as soon as you start theorizing and try to answer the question "why," the element of probability will enter your writing, as also will conjecture and persuasion.

The description of a theory can start out like the description of a machine or process. But *importance* will be the key word. The reader will want to know why it is necessary for him to be given such basic and fundamental material. He will need to know the *application* of the theory to the main purpose of the report you are writing; and it may also be necessary to include something on *historical development* so that he can understand the theory in its proper perspective.

The main part of the description of a theory is usually arranged according to one of three plans, or combinations:

(1) A new theory may call for an arrangement based on *synthesis*—the assembling of facts one by one until they can be combined into a general, all-embracing statement. This arrangement is useful when you wish to lead the reader along the path that you yourself took in reaching a conclusion.

(2) For the clarification of a complicated idea, the *analytical* method may be useful. This approach is suitable when you find it practical to break a general statement into parts and to examine each of these in detail.

(3) For a reader who is almost totally unfamiliar with the material, a theory will be clarified by *comparison* or *analogy*. With this method you can describe new theories by relating them to others generally well known.

The following short article by Albert Einstein is a good example of how a theory or scientific idea should be described. In it are found a building-up process that is logical and clear, a historical survey, and a summary paragraph using analogy.

$$E=mc^2$$

In order to understand the law of the equivalence of mass and energy, we must go back to two conservation or "balance" principles which, independent of each other, held a high place in pre-relativity physics. These were the principle of the conservation of energy and the principle of the conservation of mass. The first of these, advanced by Leibnitz as long ago as the seventeenth century, was developed in the nineteenth century essentially as a corollary of a principle of mechanics.

Consider, for example, a pendulum whose mass swings back and forth between the points A and B. At these points the mass m is higher by the amount h than it is at C, the lowest point of the path (see drawing). At C, on the other hand, the lifting height has disappeared and instead of it the mass has a velocity v. It is as though the lifting height could be converted entirely into velocity, and vice versa. The exact relation would be expressed as $mgh=\frac{m}{2}v^2$, with g representing the acceleration of gravity. What is interesting here is that this relation is independent of both the length of the pendulum and the form of the path through which the mass moves.

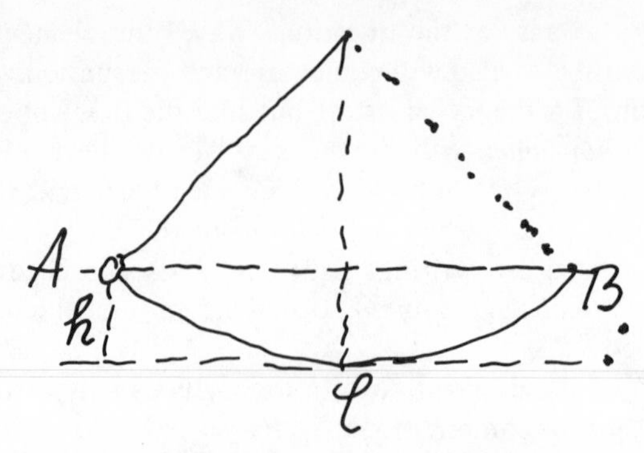

Drawing from Dr. Einstein's Manuscript

The significance is that something remains constant throughout the process, and that something is energy. At A and at B it is an energy of position, or "potential" energy; at C it is an energy of motion, or "kinetic" energy. If this concept is correct, then the sum $mgh + m\frac{v^2}{2}$ must have the same value for any position of the pendulum, if h is understood to represent the height above C, and v the velocity at that point in the pendulum's path. And such is found to be actually the case. The generalization of this principle gives us the law of the conservation of mechanical energy. But what happens when friction stops the pendulum?

The answer to that was found in the study of heat phenomena. This study, based on the assumption that heat is an indestructible substance which flows from a warmer to a colder object, seemed to give us a principle of the "conservation of heat." On the other hand, from time immemorial it has been known that heat could be produced by friction, as in the fire-making drills of the Indians. The physicists were for long unable to account for this kind of heat "production." Their difficulties were overcome only when it was successfully established that, for any given amount of heat produced by friction, an exactly proportional amount of energy had to be expended. Thus did we arrive at a principle of the "equivalence of work and heat." With our pendulum, for example, mechanical energy is gradually converted by friction into heat.

In such fashion the principles of the conservation of mechanical and thermal energies were merged into one. The physicists were thereupon persuaded that the conservation principle could be further extended to take in chemical and electromagnetic processes—in short, could be applied to all fields. It appeared that in our physical system there was a sum total of energies that remained constant through all changes that might occur.

Now for the principle of the conservation of mass. Mass is defined by the resistance that a body opposes to its acceleration (inert mass). It is also measured by the weight of the body (heavy mass). That these two radically different definitions lead to the same value for the mass of a body is, in itself, an astonishing fact. According to the principle —namely, that masses remain unchanged under any physical or chemical changes—the mass appeared to be the essential (because unvarying) quality of matter. Heating, melting, vaporization, or combining into chemical compounds would not change the total mass.

Physicists accepted this principle up to a few decades ago. But it proved inadequate in the face of the special theory of relativity. It was therefore merged with the energy principle—just as, about 60 years before, the principle of the conservation of mechanical energy had been combined with the principle of the conservation of heat. We might say that the principle of the conservation of energy, having previously swallowed up that of the conservation of heat, now proceeded to swallow that of the conservation of mass—and holds the field alone.

It is customary to express the equivalence of mass and energy (though somewhat inexactly) by the formula $E = mc^2$, in which c represents the velocity of light, about 186,000 miles per second; E is the energy that is contained in a stationary body; m is its mass. The energy that belongs to the mass m is equal to this mass, multiplied by the square of the enormous speed of light—which is to say, a vast amount of energy for every unit of mass.

But if every gram of material contains this tremendous energy, why did it go so long unnoticed? The answer is simple enough: so long as none of the energy is given off externally, it cannot be observed. It is as though a man who is fabulously rich should never spend or give away a cent; no one could tell how rich he was.

Now we can reverse the relation and say that an increase of E in the amount of energy must be accompanied by an increase of $\frac{E}{c^2}$ in the mass. I can easily supply energy to the mass—for instance, if I heat it by 10 degrees. So why not measure the mass increase, or weight increase, connected with this change? The trouble here is that in the mass increase the enormous factor c^2 occurs in the denominator of the fraction. In such a case the increase is too small to be measured directly, even with the most sensitive balance.

For a mass increase to be measurable, the change of energy per mass unit must be enormously large. We know of only one sphere in which such amounts of energy per mass unit are released: namely, radioactive disintegration. Schematically, the process goes like this: An atom of the mass M splits into two atoms of the mass M' and M'', which separate with tremendous kinetic energy. If we imagine these two masses as brought to rest—that is, if we take this energy of motion from them—then, considered together, they are essentially poorer in energy than was the original atom. According to the equivalence principle, the mass sum $M' + M''$ of the disintegration products must also be somewhat smaller than the original mass M of the disintegrating atom—in contradiction to the old principle of the conservation of mass. The relative difference of the two is on the order of 1/10 of one per cent.

Now, we cannot actually weigh the atoms individually. However, there are indirect methods for measuring their weights exactly. We can likewise determine the kinetic energies that are transferred to the disintegration products M′ and M″. Thus it has become possible to test and confirm the equivalence formula. Also, the law permits us to calculate in advance, from precisely determined atom weights, just how much energy will be released with any atom disintegration we have in mind. The law says nothing, of course, as to whether—or how—the disintegration reaction can be brought about.

What takes place can be illustrated with the help of our rich man. The atom M is a rich miser who, during his life, gives away no money (*energy*). But in his will he bequeaths his fortune to his sons M′ and M″, on condition that they give to the community a small amount, less than one thousandth of the whole estate (*energy or mass*). The sons together have somewhat less than the father had (*the mass sum* $M' + M''$ *is somewhat smaller than the mass M of the radioactive atom*). But the part given to the community, though relatively small, is still so enormously large (*considered as kinetic energy*) that it brings with it a great threat of evil. Averting that threat has become the most urgent problem of our time.

—*Science Illustrated*

4.4 Variations in Description

As with all kinds of writing, the style and complexity of technical description depends on the reader for whom you are writing (see Sec. 2.3). Therefore, you may have to consider variations in the techniques we have already suggested.

In describing a machine for a layman, for example, it may be desirable to start out by explaining the ways in which the machine affects him. He may be interested in efficiency, in the newness of the device, or in its uniqueness. In other words, be particularly strong in *reader motivation* and personal appeal.

For such a reader, we should like to stress again the use of *analogy*. Analogy is an informal type of definition, using comparison and contrast. We recall reading an article on aircraft hydraulic systems in which, to quote the author, "The unloading valve in the pump acts like a policeman directing traffic." And in another place the author says: "In a basic hydraulic system it should be remembered that the pump is the most important part, just as the heart is the most important part of the human system."

For the layman, you may have to use *dramatization* and *incident*. The description of a process in celestial navigation was most effectively pointed up by taking an imaginary character, John Doe, through the ordeal of being set adrift in mid-ocean and showing how he put to practical use his knowledge of celestial navigation.

The theory is sometimes complicated by the fact that you must not only describe it—you must persuade the reader to accept it. Here is where two principles of argumentation may enter your writing: (1) at the outset establish a framework for your arguments, defining the premise on which your arguments will be based and the *limits* you will set; and (2) *anticipate objections* that may arise in the mind of your reader. These objections are bound to appear if your subject is at all controversial; either admit their validity or prove their falsity before they become fixed in your reader's mind.

4.5 Suggested Outlines

The outlines below sum up our suggestions for basic descriptions of machines, processes, and theories:

Machines

(1) Nature and Function
(2) Theories and General Principles
(3) General Description
(4) Specific Description
 —by order of function
 —by order of importance
 —by order of spatial arrangement

Processes

(1) Nature and Function
(2) Theories and General Principles
(3) Materials and Machines Involved
(4) General Procedure
(5) Detailed Steps in the Process
(6) Variations in the Process
(7) End Result

Theories and Ideas

(1) Nature and Function
(2) Importance of the Idea
(3) Application
(4) Historical Development
(5) Description by
 —synthesis
 —analysis
 —comparison
(6) Summary

EXERCISES

1. The design for a new automobile seat has been released to the technical press. As part of an article on highway safety, describe it for a technical reader. The data are:

Trade name: Survival Capsule Seat.
Form-fitting seat of the usual size.
Provides a protection during rear-end collisions by preventing direct intrusion forces from reaching motorist.
Includes such safety features as a roll bar, built-in full harness restraint, padded side impact wings, padded armrests, and padded head supports.
Resists intrusion during side-impact encroachment by isolating motorist from direct blow of striking car.
Tested at Institute of Transportation at Wrigley, Long Island.
Revolves to facilitate entrance and exit.
Moves forward and backward to accommodate height of driver.
Titanium-alloy tubing provides high strength with light weight.
Sits on pedestal.
Built-in full harness restraint. Head support resists roof collapse.
Angled at 75 degrees.
Seat to floor, 10 inches; seat to shoulders, 22 inches; shoulders to top of head, 8.5 inches.
Envelops motorist at sides and rear.
Tests now being carried out.

2. A new machine called the Secret Recorder has recently been released to the public. As part of a manual of operation, the machine must be described. The data are as follows:

Trade name: Secret Recorder.
Manufactured by a well-established and dignified electronics corporation.
Product comes camouflaged in an average-sized leather briefcase, which may be opened, carried, or put down without revealing the fact that the recorder is in operation.
Dimensions: 16 x 12½ x 4½ inches.
Recording microphone is built in.
Unit operates entirely from batteries: the motor from five mercury batteries, the amplifier from standard dry cells.
There is a built-in preamplifier for headphone playback.
Unit records continuously for 90 minutes at a tape speed of 1⅞ inches per second, using long-play tape.
A 5-inch reel, therefore, accommodates 3 hours of dual-tracking recording.
Recorded tape may be played back on any standard machine operating at 1⅞ inches per second.
Rewinds electrically in 2 minutes.
Unit is in full-scale production.
Fidelity is high considering the small size.
Weight: 11¾ pounds.
Manufacturer indicates that the product is excellent for investigative work.

Describe the machine for the consumer, assuming that a manual of use will follow your instructions.

3. The Palmer Casting Corporation of Fletcher, California, has produced an unusual steel shaft to be used in a new wind tunnel being built at the Burbank Engineering Center, Telemanee, Wisconsin.

You have been given the following information:

Steel shaft connects wind-tunnel motors to compressors of the transonic tunnel.
Weight: 50 tons.
Length: 25 feet.
Diameter: 2.5 feet.
Diameter of coupling: 5.5 feet.
Power carried from four motors (two of these larger than any previously built for the purpose).
Vibration: less than 0.001 inch.
Shaft floated in oil on bearings.
Can be turned by one hand pushing on brake wheel.

Write a description of the piece of machinery, to be used in a technical magazine.

4. An electronics company has designed a machine called a 604 Magnetic Tape Recorder. As part of a manual of operation, the machine must be described. The data are as follows:

A high-speed input-output unit designed to provide storage capabilities required by company's computers and data-processing systems.
Tape unit is 72 feet high, 33 feet deep, 28 feet wide.
Bank of switches and indicators on front control panel allows monitoring and manually controlled tape operations.
Bank of switches and indicators on rear maintenance panel provides operator control during test procedures.
Information stored on magnetic tape in the form of magnetized tapes or bits.
Weight: 1200 pounds.
Logic control section mounted on the rear of cabinet.
Each tape unit includes self-contained read/write electronics, control logic, and power supplies.
The 604 may be used with computers in an on-line capacity or with external equipment in an off-line processing system.
Power input provided by 60-cycle, 208-volt, 3-phase, 15-ampere service line.
Tape unit located in front of transport; maintenance panel at rear.
Each 604 operates independently of other units in the system.
Logic section consists of a number of circuits mounted on separate, pluggable, printed circuit cards.
Supply reel and take-up reel provided.

Describe the mechanism for the consumer, assuming that a section on operation will follow.

5. For a manual of operation, it is often necessary to describe a machine or device before you tell someone how to use it. In 200 words or less, describe such a machine, assuming that a set of instructions will follow.

6. A student working in a laboratory may be confused by the equipment he must use. Describe a piece of laboratory test equipment such as a theodolite, a hardness-testing machine, or a cathode-ray oscilloscope. Assume that the description will be included in a laboratory manual.

7. Assume that you manufacture a machine for commercial use, such as a tape recorder, a slide projector, or a Mimeograph machine. Your salesmen require a description of the machine for demonstration purposes. Write such a description for this specific group.

8. The following paragraph is the beginning of a short description of basic oxygen steelmaking.

> America's capability to produce steel by the basic oxygen process has grown enormously from small beginnings during the middle 1950's. The high tonnage of steel now made in basic oxygen furnaces —commonly called BOF's—requires the consumption of large amounts of oxygen to provide operational heat and to promote the necessary chemical changes. No other gases or fuels are used.
>
> The basic oxygen process produces steel very quickly compared with the other major methods now in use. For example, a BOF may produce up to 300-ton batches in 45 minutes as against 5 to 8 hours for the older hearth process. Most grades of steel can be produced in the refractory-lined, pear-shaped furnaces.

Complete the description of the process for a technical magazine by using the resources of your college library.

9. Describe a process as it occurs in nature. Examples of such a process are an earthquake, a tidal wave, migratory flights of birds, a thunderstorm. Write the process first for technical readers; then adapt it for semitechnical readers.

10. Assume that you have been asked to write portions of a high school text on general science. Your contribution in each case is to be the description of a theory that is to be followed by a section on practical applications and demonstrations. Write the theory section for one item, using a maximum of 300 words.

11. Here is a suggested list of topics to be used in exercises in technical description.

Machines and Mechanisms
- electric typewriter
- Thermofax machine

x-ray microscope
wind tunnel
automobile engine
solar cell
hydrofoil

Processes and Procedures

xerography
color printing
student registration by computer
scatter communication system
irradiation of foods
programmed instruction

Theories

operations research
cybernetics
heat transfer
solid-state physics

II

The Report

5. Reports: General

5.1 Types of Reports

Some textbooks on technical writing divide written reports into a large number of carefully distinguished classifications. They may list as distinct types information reports, periodic reports, progress reports, examination reports, and recommendation reports. They may separate letters into letters of inquiry, answers to inquiries, letters of instruction, and letters of transmittal. They may make distinctions between descriptive reports, analytic reports, evaluative reports, and many others.

You will notice that these listings classify the reports by their subject matter rather than by their form. Now we believe that in general all technical reports have the same functions to fulfill and that they are governed by the same principles no matter what their subject. For instance, suppose you were writing a report on the metallurgical analysis of a failed casting. Perhaps you were told simply to determine the cause of failure, and stop there. Your report would then be an "analytical" report. On the other hand, you might have been told to make recommendations about foundry practice to eliminate such failures in future castings. In this case, your report would be a "recommendation" report.

But how would these two reports differ? Only in the presence or absence of the recommendations. Would they be essentially different? Would your general approach and your problems in writing them be any different? We believe not.

Therefore in *Technical Reporting,* except in connection with a few short reports, we will not classify reports by their subject matter, but by their form; and the two broad classifications we will make are overlapping and far from rigid. They are (1) the long or formal report and (2) the short or informal report.

Whether a report is long or short, it should fulfill the requirements discussed in Chapter 2: it should orient the reader and provide him with a framework on which to hang the parts, and then it should transmit its information to him clearly and efficiently. Whether it is long or short, it will do these things better if it has been prepared in accordance with the principles presented in Chapter 2 and Chapters 14–19.

But the longer a report, the more it needs (1) a sturdy and convenient binding and (2) an obvious, formal organization, with its parts clearly labeled and listed. It is these two comparatively superficial differences that distinguish the formal report from the informal report. The formal report (Chapter 7) is usually bound in a cardboard cover, or perhaps even in a cloth binding. It is usually divided into separate main sections, each beginning on a new page, and it has a table of contents. The informal report (Chapter 6) is usually simply stapled together, with no cover. It is likely to be loosely divided by means of informal subheadings rather than separated into major sections, and if it is short it can get along without a table of contents. But some informal reports are quite highly organized, with forewords, summaries, tables of contents, and other parts that are usually associated with more formal documents. Thus the formal report may sometimes be differentiated from the informal only by the presence of a cover and a title page.

The length criterion does not establish a line of demarcation between formal and informal reports, either. Ten pages might serve as an arbitrary dividing line between the long or formal report and the short or informal report. Yet we have seen formal, bound reports with no more than three pages of text, and informal, unbound reports of more than fifty pages. The decision whether to make a report formal or informal depends not only on its length, but also on the readers for whom it is intended, and how much of an impression you wish to make on them; the finality of the judgments in it; and the time available for its preparation.

So far we have mentioned only *reports,* formal and informal (letters are included under informal reports in Chapter 6). In addition, *Technical Reporting* discusses four other sorts of communications—instructions in Chapter 10, proposals in Chapter 11, technical papers and articles in Chapter 12, and oral reports in Chapter 13.

5.2 Forms of Reports

In Chapters 6 and 7 you will find some suggestions for the form, or makeup, of various kinds of reports and letters and their headings, and in Appendix Sec. 21.2 you will find reproductions of reports and letters that illustrate them. These forms have actually been used by successful organizations, and they are clear and effective. *But we wish to emphasize that they are simply examples from a wide choice of acceptable forms. The organization for which you work will very likely have its own standardized forms and headings for various written communications. Unless you can convince your superiors that these forms should be changed, you will of course abide by the local rules, even though they differ from the suggestions in this book.*

6. Informal Reports

6.1 General Principles

Chapter 5 pointed out some of the ways in which informal reports differ from formal reports. In general, informal reports are less finished—less polished—than formal reports; they are appropriate for informal situations. Although we cannot make a clear-cut distinction, informal reports are more likely to be *internal* reports; that is, the readers are probably within the company itself. Informal reports are not written for wide distribution.

For our purposes we can divide informal reports into two general classifications: (1) the short-form report or memo report, discussed in Sec. 6.2, and (2) the letter report, discussed in Sec. 6.3.

The informal report—particularly if it is short—does not need the explicit formal organization of the formal report. But reports of all kinds have essentially the same job to do, so that they need the same functional elements if not the same outward form. Thus three features regularly associated with formal reports will usually help even the short informal report to transmit information efficiently:

(1) An early statement of the problem or purpose. Even when a man is generally familiar with a situation, he needs some introduction—a lead-in—before he is ready to plunge into it. Your boss may know that he has told you to investigate the weight of claw hammers; but he has a lot of other matters

on his mind, too, and a lot of other people are investigating other things for him. Therefore, before you tell him the weight of claw hammers, tell him that you are about to report on the weight of claw hammers. This principle is illustrated in the report reproduced in Sec. 21.2.1.

(2) An early statement of important results or conclusions. It is now standard practice to state the most important results and conclusions near the beginning of long formal reports (see Chapter 7). Although many people will disagree with us, we believe that short informal reports, too, are usually improved by an early, unsupported statement of the important information being reported. Your boss may have enough confidence in you to take your word on the weight of claw hammers without reading about how you got your data. If he does, you can save him time by giving him the data right away, as soon as you have oriented him. This method is used in the report in Sec. 21.2.8.

(3) A liberal use of heads and subheads. The longer an exposition, the more it needs to be broken up by headings. But a liberal use of headings can often make even a short report clearer and more useful. The short memo reports in Secs. 21.2.1, 21.2.2, and 21.2.3 are all clarified by their subheadings.

6.2 Memo Reports

6.2.1 Use of Memo Reports

A kind of informal report that is useful in a wide variety of situations is known as the memorandum report, or memo report. Although it is often sent to outsiders, the memo report is particularly suitable for the reporting of technical information within the issuing organization. Three kinds of subjects lend themselves particularly to memo reports: trivial subjects, subjects of only temporary interest, and more important subjects that must be taken care of immediately. For example, the preparation of a formal report takes time; you may be called upon to write a preliminary report before you have had a chance to think much about a formal report on the same subject. Progress reports, for example, are often in memo-report form.

A memo report, then, may be used whenever a formal report is not warranted by the length, the importance, or the finality of the subject, or by the nature of the audience for which it is written.

6.2.2 Form of Memo Reports

Because of the way they are used, memo reports should be short and to the point. Memo reports as they are now used in companies have become largely standardized both in form and organization. However, you may modify any standard form to suit your own writing situations within the framework of company requirements.

Memo reports do not customarily have a separate cover or a separate title page. The text begins a few lines under the heading, which appears at the top of the first page. The heading supplies appropriate information from among the following:

Name and address of issuing organization
Report number
Title
Name of person to whom report is addressed
Name of author
Date
Number of pages

6.2.3 Organization of Memo Reports

Each memo report should deal with only one subject. Sometimes, because this kind of report is so short, you may be tempted to string several memoranda together. You will only confuse your reader. Write a new memo for each subject.

The following organization is suggested for the memo report.

(1) *Subject line.* You may find that the title is not enough. Use a subject line containing more information. Put it in all capitals at the beginning of the report proper to draw attention to the content and to make filing easy.

(2) *Statement of the situation or procedure leading to the subject under consideration.* At this point, make clear reference to any link with past work.

(3) *Conclusion, recommendations, or disposal of the problem.* In a sense, this is the meat of the discussion and the part in which the reader is particularly interested. Treat it as fully as need be, but remember that you are writing a memo report.

(4) If necessary, detailed explanation of the thinking or procedure which led to the statement in 3.

Various names are given to these parts. The material in 2 may be called the *foreword;* 4 may be called *discussion.* Different orders are often used. For example, if the memo is particularly short, the discussion in 4 may precede the conclusions in 3. In general, the longer the memo, the more useful it is to have the conclusions stated early.

Examples of acceptable memo-report forms appear in Secs. 21.2.1–21.2.7.

6.2.4 Service Reports in Memo Form

You will find the memo form particularly useful when you have to write service reports. Service reports help a company carry on the everyday transmission of technical information. Sometimes they are ends in themselves; for example, a department must write a trouble report when a piece of machinery fails. This trouble report is important in itself; it provides information for the repairman who fixes whatever went wrong. At other times service reports add another dimension; they become the means by which another publication is written, whether it is to be another report or a manual of some kind. An example is the progress report. When you write a progress report, you will not only be telling about something that has happened in the immediate past, but you will probably use the report and others like it when you come to write a final report.

6.2.5 Kinds of Service Reports

Most companies use a great many service reports. They go by various names; some are common to all companies, others are used only by individual companies.

In general, however, report writers write the following service reports at one time or another: (1) the survey report, (2) the progress report, (3) the status report, (4) the trip report, (5) the conference report, and (6) the trouble report. If you are still in college, you may have occasion to write some of these reports. (We hope that you won't have to write a trouble report.)

Sec. 6.1 pointed out the three features regularly associated with all kinds of reports: an early statement of the problem or purpose, an early statement of important results or conclusions, and a liberal use of heads and subheads.

Service reports will incorporate these features, too, but here are some further suggestions to add to those already given.

(1) Survey Report

You may be asked to write a survey report before starting a new project, say a piece of research. If your company is expanding, for example, a survey will be made of new markets, new locations, and other new factors. Or another company may make a proposal and try to sell you something; you may find it necessary to make a survey of certain parts of your company to see if a change is desirable.

The survey report should start with a clear statement of why the survey was undertaken. Then, because the reader will want to see your conclusions and recommendations as soon as possible, these should follow immediately. But surveys are frequently capable of varying interpretations. In a survey report, especially, you should tell what you did, what you saw, what data was available to you. More often than not, you will find it necessary to record oral presentation of statements from other people. Estimated costs and a forecast of future events may be included.

Specimen survey reports appear in Secs. 21.2.3 and 21.2.4.

(2) Progress Report

Progress reports are issued at intervals to show what has been done, what is being done, and what is expected to be done. When several people are working on one project, one report often covers the work of all of them.

Progress reports should be made on the basis of a definite amount of work accomplished. They may be made daily, weekly, or at other designated intervals.

A progress report starts out with a statement showing how much ground the report covers, and, unless it is a first report, contains information linking it with former reports. Then may follow a brief statement of the work that had been covered by past reports. This is not needless repetition; it is necessary to show the continuity of work.

Then comes a discussion of the new work being reported. As this is the principal part of the progress report, it should be discussed completely within any limitations of space previously agreed on. Sometimes progress reports contain recommendations for changes in procedure or new courses of action. Just as frequently they do not. They usually end, however, with a prediction of how much will be accomplished in the future, or how much remains to be done.

A specimen progress report is reproduced in Sec. 21.2.5.

(*3*) *The Status Report*

The status report is a special kind of progress report. It may integrate the work of a number of departments or a number of projects, at certain stated times. It will evaluate not only the work being done but also the accomplishments of the personnel involved. For example, in a research laboratory it may not be possible to accomplish anything very tangible or practical over a period of time. Yet it is necessary to know what the research people have been doing. Because the status report usually covers more ground than the progress report, it is likely to be less specific.

(*4*) *Trip Report*

Trip reports are usually the record of business trips made and the information gathered. They describe observations and conversations. They contain some of the features of the survey report, but always include a complete list of people seen and digests of conversations and conferences. Because conversations are likely to become repetitious and to digress from the subject, the writer of the trip report must make a special effort to sort out information, to use headings liberally, and to be very careful about where emphasis is placed. Conclusions and recommendations may be included in the trip report, but usually the emphasis is on what actually was covered at the time the trip was made. Places, times, and dates should be accurately set down. Specimen trip reports appear in Secs. 21.2.6 and 21.2.7.

(*5*) *Conference Report*

Many engineers and other professional people spend a considerable amount of time attending conferences. In order to justify the time and money spent on going to conferences and to bring back to their companies and agencies the information acquired, they usually write another kind of trip report called a conference report. The conference report will give the time and place of the conference, the reason for it, and the results achieved.

Some conferences, more general than others, are planned to appeal to a variety of interests, such as those of professional societies. Each conference features a general theme to which it tries to adhere. For example, a recent conference of the Society of Technical Writers and Publishers was built around the theme *Technical Communications—Man's Record of Reality*.

Other conferences are intracompany; that is, they bring together divisions within a company or government agency. For example, the New York State Department of Health has branches in Buffalo, Albany, and Rochester, to name a few. Each participant at a conference might be expected to write a report when he returns to home base. The material for a report of this kind will be adapted from the program of the conference and should describe not only the speakers and the theme of the speeches, but also any overall conclusions reached.

(6) *Trouble Report*

Even in the best-regulated families, troubles will occur. In college, you will find that laboratory experiments go wrong; you may come up with negative results because of failures of apparatus or failure to follow directions. In industry, defects occur—defects of machinery, performance, and servicing.

The trouble report is usually written for a closed circle of readers. It states the reason for the report's being written in the first place, the situation as it was found by the writer, what was done to correct the situation, and how successful the solution was. There usually follows a clear and careful analysis of the factors leading to the trouble. Often this analysis involves personalities. It is well to state the facts definitely, but impersonally.

The trouble report usually ends with certain recommendations or suggestions. This section may not be found in the beginning, as in many other kinds of reports, because in the trouble report emphasis is on the immediate correction of the trouble. But remember that trouble reports serve the dual purpose of supplying an immediate record and a method of changing and improving existing situations.

6.3 Letters

Technical information is often transmitted in letter form. However, such a letter is really an informal report. The principles of technical reporting presented in this book apply as well to letters as to other kinds of reports.

In any letter, whether a letter report or a business letter, certain principles should be followed:

(1) Establish contact with your reader and orient him before plunging into the subject.

(2) Say everything in a straightforward, matter-of-fact way. You should not try to be particularly ingratiating or overfriendly, nor should you be

brusque or in any way rude. Use the simple, everyday language that you use in other reports (see Sec. 14.2).

(3) Use the letter as an opportunity for being more personal than in any other kind of report.

(4) Use headings and subheadings wherever you think they will clarify the meaning. They can be helpful, particularly in a long letter or one that treats several subjects (see Sec. 21.2.4).

(5) Present the subject as much from the reader's point of view as possible. In persuasive letters particularly, appeal to his interests as you try to get him involved in your problem.

The form of the letter has become fairly standardized. The specimen letters in Secs. 21.2.8–21.2.14 are typical.

6.3.1 Letter Reports

When you go into the engineering department of a company and letters are mentioned, you will usually find that it is the letter report that is being referred to. The letter report combines the features of the formal business letter and the short informal report. It differs from the straight report in that it is usually addressed to one person rather than to a group. That person may be within the company or outside it. For example, an engineer may write to an executive in the management; or he may write to his opposite number in another company. When it is addressed to a person, the letter report frequently displays the informality found in letters but not generally found in reports. It can be written in a positive, personal style rather than in the impersonal, objective style of the report.

Like any other informal report, the letter report illustrates the features cited in Sec. 6.1. It usually has a structure that gives the reader an early statement of the problem or purpose; an early statement of important results and conclusions; and a liberal number of headings and subheadings. In fact, these three features of the report are becoming identified with a variety of business letters. A specimen letter report is reproduced in Sec. 21.2.8.

6.3.2 Business Letters

It is difficult to say where technical information stops and nontechnical information begins. The nontechnical letter, however, is usually called a *business letter,* and will differ from the letter report by being even more informal in style. It will be addressed to a reader outside the issuing company.

Business letters presenting nontechnical information can be divided into two types, according to their uses. One type is the *informative* business letter, which provides routine information, perhaps in reply to a request.

The second type of business letter is *persuasive*. Suppose you are asking a firm for material to be used in writing a thesis; you may find it necessary first to sell the firm the idea that you need the data, even before you make the request. A business letter of this kind takes on the form of the *sales* letter with its classic techniques of attracting the attention of the reader, creating interest in the idea or product, convincing the reader that the information you need is necessary, and persuading him to do something for you.

See Sec. 21.2.12 for an example of a business letter.

6.4 Job-application Letters

6.4.1 General Principles

A special kind of letter—and an important one—that most people have to write sooner or later is the job-application letter. The application letter can very appropriately be thought of as a combination report and sales letter.

As a report, the letter gives information about you. Again, the principles of good reporting apply: tell your story as accurately, as clearly, as simply as you can, and be as brief as possible without omitting any important information.

As a sales letter, the job-application letter must persuade the reader to read your letter, to accept your information, and to come to some decision about you. You can therefore borrow from what are considered the classic techniques of the sales letter (see Sec. 11.3).

We will show you how this works out. For example, some people have trouble in beginning an application letter, in using devices that will make the recipient of the letter want to read it. The first part of a sales letter attempts to attract attention. So should the application letter.

The attention of the reader can be drawn to the letter in various ways. One of these is purely mechanical. Write a letter that *looks* easy to read. It should be well typed, it should follow the traditional form of the business letter, and it should contain reasonably short paragraphs and sentences. Some of the material may be in tabular form so that certain items stand out.

Another device attracts attention through what is called *identification*. Make it easy for the reader to find out who you are and what you want. You can use a report technique: tell him at the beginning what your letter is about. That is, say simply that you are writing to apply for or inquire about a certain job. A second effective means of achieving identification is by establishing some personal relationship. If you can open the letter by referring to a mutual acquaintance, to an advertisement that appeared in a paper, or to the student placement office of your college, it establishes you within a definite context. Do not be afraid to use personal relationships; their use is a perfectly legitimate device to awaken interest.

The second part of the application letter creates an interest in you and your ability. You may be writing a solicited letter; that is, it may be in response to an advertisement, or the college placement bureau may have let you know that a company is recruiting new personnel. Thus the company is already interested in you to some extent. On the other hand, you may be attracted to a company that is not at the moment recruiting new personnel. You will then write an unsolicited letter and your task of creating an interest in yourself will be greater.

One method of creating interest is by expressing enthusiasm for the job. A desultory and lifeless letter will usually get a negative reaction. Since the letter of application is a report on yourself, you will inevitably have to use the pronoun *I*. Any effort to avoid *I* by means of third-person or passive constructions will sound artificial and stilted. (See Sec. 14.13 on impersonality.)

Some application letters we have seen throw most of the responsibility upon the employer. The writer does not seem to know what kind of job he wants or what kind of job he is qualified for, or indeed what kind of company he is applying to.

A letter of application, then, should contain in the second part all or some of these elements:

(1) A statement of the kind of job the writer is seeking. Most companies are so large and diversified that to write a general application letter is folly.

(2) Evidence of some knowledge of the company and what it produces. In large companies, reference should be made to departments or divisions.

(3) Identification of the writer through reference to his education and training. Again, it is a waste of time for the wrong person to apply to the wrong place. It is analogous to a high school student interested in engineering seeking admission to a medical school. If you aren't selective in your applications, you will be among the first group weeded out by the prospective employer.

In general, then, if you feel that you are particularly well qualified for the job, tell the employer. In an application letter, you have to blow your own horn; try to do it in a quiet, objective way that does not sound conceited.

Now that you have whipped up an interest in yourself, you must deliver the goods. This part of the application letter—convincing the reader that you are the right person for the job—takes in your education, any special training that you have had, and any experience. In the next section, we will be talking about two arrangements for application letters. But regardless of the kind, remember that you are writing a sales letter and that you are in competition with other people. The application letter must stress those things that make your education and experience more applicable to the job than those of another person's.

The curriculum you have been taking will be similar to the curriculum in the same subject in another college. Therefore there is little need to draw undue attention to it. But in engineering, for example, new directions in curricula have come into being in the last few years. If you are a graduate of a typical chemical engineering curriculum, that may be all you have to say about it. But perhaps your school has incorporated chemical engineering with biomedical engineering and has produced a number of highly specialized courses. This is the kind of thing that makes you stand out from other applicants, and it should be emphasized in your letter.

Certain courses outside your major curriculum should be cited. Many companies with foreign branches are looking for graduates with a knowledge of a foreign language. If you can speak or read any language in addition to English, by all means mention it.

All companies hope that they are attracting future managers and vice presidents, people with that extra "something." If you have taken courses that in any way enhance your market value, such as management, psychology, philosophy, English, don't hesitate to mention them. If you think that you are a particularly good writer and can put together a good report, and have the experience to prove it, be sure to mention it to the prospective employer. (By all means your letter should reveal this ability!)

Employers are also looking for the student who has put his special characteristics to good use. If you have been an officer in a campus club or a member in a professional student organization like the student branch of IEEE, be sure to mention it. Call attention also to any honors you have received, scholarships, citations for superior work, and papers you have had published in student journals or elsewhere. The employer will want to know about them.

He also likes to find out what practical experience you have had. And sometimes off-beat summer employment will attract his attention—that you washed dishes in a restaurant, for example, or that you worked on highway construction. Jobs such as these, which may be outside your profession, nevertheless show that you possess initiative, perseverance, and the ability to adjust to varying situations.

The closing part of the application letter should not seem to be an afterthought; it should end strong. Decide what your letter is supposed to accomplish and what you want the reader to do. Although the ultimate goal of the application letter is a job, the immediate goal is an interview. Because most organizations do not employ anyone without a personal interview, almost every letter should specifically request one.

Make it easy for the employer by stating the times you are available for an interview. Employers realize that students, especially seniors, are committed to rigid schedules. You are not being egotistical or demanding if you let the employer know when you can come to see him. And by all means include your telephone number somewhere in your letter. It is possible that a telephone call will be easier for the employer than a letter.

6.4.2 Form of Application Letters

Two general arrangements prevail in application letters: (1) an ordinary business letter that contains all the information being transmitted; (2) a short letter accompanied by a data sheet that contains most of the personal and historical information in tabular form. The data-sheet arrangement is now the commoner of the two. Yet the straight letter has an advantage if you are a proficient writer: it gives you a chance to demonstrate your skill at writing, a quality that most employers value highly. Examples of the two arrangements are given in Secs. 21.2.13 and 21.2.14.

If you do use data sheets, you can reproduce them in three possible ways: (1) you can type them individually; (2) you can duplicate them by Xerox or some other process; and (3) you can fill in the spaces of a prepared form such as the ones supplied by some colleges. The prepared form must be constructed to fit the needs of many diverse people. For you, it will probably have too much space for some items, not enough for others; it will probably be longer than you need. We believe you will do well to construct your own data sheet rather than use a prepared form. Here, again, you can demonstrate a skill to your prospective employer.

Although at one time it was considered disastrous to have data sheets duplicated, the fine results from Xeroxing or similar processes have about done away with that prejudice. Most graduating students canvass a large number of prospective employers. A duplicated data sheet saves time and can be as attractive as the original. Moreover, employers have come to expect it. The accompanying letter, however, should always be original for each company solicited.

6.4.3 Contents of Application Letters

In general the application letter should contain information about yourself that bears directly on your application—information that will help your prospective employer judge whether you are likely to be well fitted for the job. Below is a list of subjects that are appropriate. You may of course have reason to omit some of them or to add others, and their order is not significant. The starred items are the ones that customarily go on the data sheet if you are using that form; some of the unstarred items may also be transferred to the data sheet. The order of the items is not significant. In the data sheet, though, reference to physical condition, address, age, and citizenship are usually placed near the beginning.

Item	*Comment*
1. Type of work desired and reason	
2. Date you will be available	
3. Location desired	If you are willing to go anywhere, it may be a good idea to say so.
4. Specific reasons for wanting to work for this organization	Stress this item if you can do it honestly.
*5. Education a. Specialization b. Record	Usually given with emphasis on professional training. How far back should you go? Only to a point that bears on

Item	*Comment*
c. Honors d. Extracurricular activities	this job. Attendance at a certain high school may carry some weight; what kindergarten you went to almost certainly does not. Items dealing with your education are usually given in reverse chronological order. Opinions differ about the listing of extracurricular activities. At any rate, the purpose of mentioning them is not to show that you were a big man on campus, but to show (1) that you like to mix with people and join in group activities and (2) that you have had experience that will help you in the job you are applying for. Thus you should not go into detail about extracurricular activities unless they have direct connection with the work you expect to do or reveal something unusual about you.
*6. Practical experience	Usually given in reverse chronological order. The student may appropriately list part-time and summer jobs, to demonstrate his industry and ambition. The older man should probably not go back beyond his professional or related jobs.
*7. Physical condition	A brief statement that your health is good; or a frank description of any condition *that will handicap you in this job*.
*8. References a. Personal b. Professional	Some people say nothing about references in the initial letter of application, but furnish them later if requested. However, they are usually put in the data sheet. Ordinarily, references should emphasize people who know you in your professional life. Do not include references in answer to a "blind ad" (a want ad that does not disclose the name of the advertiser).
*9. Name; address; age; citizenship; marital status	

EXERCISES

1. A new supervisor in a government arsenal requested that each employee in his division write a letter reporting on his activities. This is what he received from one person. Rewrite it to make it a satisfactory letter report.

Mr. S. M. Acton
Red Island Arsenal

Dear Mr. Acton:

This letter is intended to introduce you to my background as an engineer at the arsenal. At present I hold the title of Methods Engineer. This title has an unusual meaning in today's age of specialization, for it means that I am a non-specialized engineer.

My work takes me into many varied areas of our organization, e.g.: sales, product design, manufacturing, machine design, plant layout, etc. I am primarily concerned with increasing the productivity of the company in the manufacturing processes. To increase productivity, it is usually necessary to change the old ways of doing things and introduce new and better methods.

In regard to writing, this changing of old ways means that I must sell the idea of change to management. This selling is done in many forms and ways but frequently reports are the basis of the change. These reports are varied as they must often be read by technical people who will question their technical aspects, they will be read by management to whom they must be clear, concise and brief but still non-technical.

There are other forms of writing that are necessary in a job of this type. For example, instructions to plant personnel, memoranda to fellow workers and supervisors, requests to other companies for information on products, etc. Therefore, I would say that my writing in industry is as varied and non-specialized as my engineering work.

Thank you for your time,

Loren R. Baker
Methods Department
Red Island Arsenal

2. The attached short inspection report was prepared by a heating engineer after examining a defective boiler originally manufactured, but not installed, by his company. This report was to be used in two ways:

(1) by the service department of his company, in estimating the cost of repairing the boiler; and

(2) by the sales department, in revising the *Installation and Operation Manual* for such boilers, to prevent further failures.

A copy of this report will go to the customer who owns the boiler.

(1) Write a critical analysis of this report, showing what is wrong with the organization, style, layout, coverage, adaptation to purpose, etc.

(2) Rewrite the report completely, applying the principles discussed in this chapter.

REPORT ON JONES CO. JOB

History

A fellow named Doe showed us around and discussed their problem. We inspected the one idle boiler.

Background

Some years ago somebody decided to eliminate feeding of water to both steam drums, and a new connection was provided to one drum only, through the hole meant for a water-column connection. This was true of both boilers. The internal feed pipes and dry pipes have been moved.

Six tubes should be replaced in the bottom row, on the side away from the feed connection. These tubes were blistered and bowed upward, and all showed signs of color, from overheating. Several tubes had failed, failures taking the form of a circumferential split in the middle of each blister.

Findings

Blistered tubes contained a hard scale, having the appearance of calcium carbonate, about ⅜ in. thick on the bottom of the tubes. The customer advised that the side on which feed-water is introduced is usually clean, with scale forming toward the bottom of the other section of the boiler.

Feed water passes through a Non-Chem water conditioner which consists of a horizontal cylinder reportedly filled with glass or copper balls.

Recommendations

(1) The immediate cause of the tube failures and bowing is scale deposit in the tubes.

(2) The reason for localized scale deposit reported by the customer is a result of poor circulation, from feeding water to only one drum. Under such an arrangement, water for the section of boiler under the other drum can be obtained only through the common mud drum.

(3) Original feedwater connections should be installed so that water is fed to both drums of each boiler through internal feed pipes. We advised that we would furnish a sketch showing this pipe.

(4) Equip the boiler with cross-over downtake circulators. These

were found on the job, having been purchased several years ago, but not installed.

(5) Water of zero hardness, free of scale-forming properties, should be fed to the boilers. We advised that we had had no previous experience with the type of water conditioner they were using, but the presence of scale in the boiler indicated that it was not accomplishing its purpose, since, regardless of the method of water feed, there should be no scale present anywhere in the boiler. We suggested that water from the conditioner be checked for a period, and if satisfactory results were not obtained, a more conventional method of treatment be resorted to, as could be recommended by any reliable feedwater consultant.

Mr. Doe requested an estimate from the Company on the cost of fixing the boiler, which they would like our Service Department to do.

This job was visited in response to a phone call from R. B. Jones, President of the Jones Bakery.

The boiler we inspected was No. 2-1419-36 model. We sold it to Jones in 1958. It was installed by Fitzgerald. It is used to heat the bakery.

Total inspection time: 7 hours
Part no. of damaged tubes: SFX 495
Date: 10-3-67, S. L. Morse & Asst.

3. Assume that the cost of your education at college is being paid by the Engineering Products Corporation (916 Seventh Avenue, North Farmington, Ohio) under its Educational Advancement Program, directed by Mr. Henry R. Matesso. By the terms of that program, your tuition is paid by the company in return for your agreement to continue working for the company for three years after your graduation. You are further required to report on your work in current courses so that the company may keep a check on your progress and may evaluate the quality of training you are receiving. You have been told that your reports should be factual rather than interpretive or evaluative. Write such a status report on your current courses.

4. You may be preparing to carry out a project, either for the course in which this book is being used or for some other course. Make a survey of the possible sources of information you will use. Address the report to an instructor and present it in memo form.

5. Write a short progress report on the situation given in Exercise 1. Put it in either memo or letter form.

6. Every so often, and especially if you have started work in your major field, your faculty adviser will want to know what success you are having with your studies, particularly with your major courses. Write such a status report.

7. The field trip is a part of many college courses. Write a trip report addressed to the department in which the field trip originated.

8. Write a letter to a firm or individual requesting information to be used in a term paper or a thesis.

9. Write a letter to a university requesting information about the possibility of your attending graduate school.

10. Write a letter to the president of one of your college activities, suggesting a change of policy in connection with the activity or proposing a new course of action.

11. You need a summer job. Write to a firm requesting such a job. Write two letters: one to accompany a data sheet, the other to contain all the information being transmitted.

7. Formal Reports

7.1 Form of Formal Reports

Any time you have to extract information from a number of different reports, you will find your task easier if the reports all follow a uniform pattern. And of course it is easier to *write* a report that follows a ready-made plan. Certainly it is a good idea for the reports of any organization all to have the same basic, overall arrangement of major sections and a uniform system of typography and mechanics. Some organizations, indeed, use the same rigid, standardized outline for every report.

But there is a growing tendency to let the outline be general and flexible so that within the basic framework each report can be constructed to fit its own subject matter, its own purpose, its own audience. *The basic outline suggested in this chapter is thus intended to be completely general and flexible. Furthermore, a number of the items in it are intended to name functions that must be taken care of rather than to be titles for sections of the report.* Within the limits of local conventions, this outline (next page) conforms with the prevailing current practice in technical reporting.

As indicated by the dotted arrows, the summary sometimes precedes the foreword; and if a letter of transmittal is used instead of a foreword, it usually comes right after the title page.

Cover
Title page
Table of Contents
Foreword or Letter of Transmittal
Summary
Body
Appendix

The rest of this chapter discusses the various parts of the formal report. They are taken up not in their order in the report, but in the order in which you will usually work on them. Two short formal reports and portions of a longer one are reproduced in the Appendix (Secs. 21.2.15, 21.2.16, and 21.2.17) to illustrate the discussion.

7.2 Body

7.2.1 Functions; Divisions

What we have chosen to call simply the *body* of the formal report actually performs a number of functions and is usually divided into a number of parts or subsections. It is the longest and most important section of the report; all other sections are built around it—are auxiliary to it. The body tells your whole story, and it is usually complete and self-sufficient, so that the reader can get the story without reading other sections of the report, with the possible exception of the foreword. (As pointed out in Sec. 7.5, the body may depend on the foreword for some background information, as in the specimen report of Sec. 21.2.17.)

The body does some or all of these things:

(1) It states the purpose of the investigation and provides the reader with enough background information so that he will be able to understand all phases of the report (let us call this function *introduction*).

(2) It tells what you did (*description*).

(3) It tells what you found out (*results*).

(4) It analyzes, interprets, and discusses these results (*discussion*).

(5) It lists *conclusions* and perhaps *recommendations* that follow logically from the discussion (if they are appropriate and if they have been requested).

In some organizations it has been customary to divide the body into separate sections corresponding to these functions. For a report on experimental work, for example, this might produce separate sections called *Introduction, Equipment, Procedure, Results, Discussion of Results,* and *Conclusions and Recommendations.* A slavish adherence to a conventional outline like this often results in (1) a lot of repetition and backtracking and (2) a burial of the most important information under a lot of details, some of which may be routine and of distinctly secondary interest. It is usually preferable to disregard these divisions and to run the parts together in whatever way is natural, logical, and efficient. In most reports other breakdowns will suggest themselves, usually according to divisions of the investigation or of the work being reported.

Remember, then, that the titles of Secs. 7.2.2–7.2.4 name *functions* rather than sections of a report.

7.2.2 Introduction

The need for orienting the reader means that the introduction function must be looked to early; but this does not necessitate a separate section labeled *Introduction* at the beginning of every report. Sometimes the introduction information can be provided once and for all, right at the start, as in the specimen of Sec. 21.2.15. More often this kind of information is needed also at the beginning of each subsection, or perhaps from time to time within subsections. Wherever it goes, the information is highly important. Be sure you take care of it in one place or another.

The part (or parts) of the report that we are for convenience calling *introduction* states the purpose of the investigation and describes the basic scheme of the procedure or methods used. It orients the reader by supplying as much historical background as necessary and then describing the present problem. It may define the scope of the study, discussing limitations or qualifications.

Sometimes a table of definitions of terms or symbols is included in the introduction. The introduction may be a good place to define important specialized terms that are going to be used frequently in the report, but the appendix (see Sec. 7.4) is usually a better place for a table of definitions.

The introduction may describe the organization of the report, so that the reader will know what to expect, or the table of contents (see Sec. 7.6) may provide a sufficient outline.

7.2.3 Description of Procedures and Methods

In order to understand the results of your investigation, your reader must know in general what you did. Certainly you must describe the basic nature of your study or investigation very early in the discussion. But in many technical reports the significant information—the results and the conclusions they lead to—get buried under needlessly detailed descriptions of the methods used. The description of procedures in the body of the report should be no more detailed than is necessary to convey your message.

The basic description of method may sometimes be given once and for all near the beginning of the report, or it may be interspersed among the related parts of the presentation and discussion of results.

Let us say that you have taken care of the basic description, in one way or the other. The full, detailed description may then be treated in four different ways (or combinations of them):

(1) It may be put near the beginning of the report.

(2) It may be interspersed among the related parts of the presentation and discussion of results.

(3) It may be put in the appendix.

(4) It may be omitted from the report entirely, and simply kept on file in the laboratory notebook or company library.

The first two of these alternatives are appropriate when a full description of method is an integral or essential part of the main story—that is, when the primary purpose of the report is to describe a new method or process that has been developed, or when a new or unusual method has been used to obtain quantitative results. Because it describes an investigation that was not routine, the specimen report in Sec. 21.2.15 provides a full description of method before it presents results.

But when the primary purpose of the report is to present the results of an investigation, and the methods used are familiar or routine, then the detailed description is not an integral part of the story, and its presence early in the report simply obscures more important material. Particularly annoying is a prominent list of standard instruments and apparatus.

If the methods and equipment are well known to the readers of the report, it may be sufficient simply to mention them by name and to omit the detailed description from the report. Anyone who wants to repeat the procedure exactly can then get the needed details from the laboratory files. This scheme has been followed in the specimen report in Sec. 21.2.17.

On the other hand, it is often necessary to supply detailed descriptions of even routine procedures for the sake of a minority of readers, or for legal reasons, or for record purposes, or to conform with company rules. You can fulfill these obligations by putting detailed information in the appendix and referring to it at an appropriate point in the text.

In short, it is a good idea to provide only as much information about methods and apparatus as your readers will need.

7.2.4 Results; Discussion; Conclusions and Recommendations

Many reports present results in one section and then discuss them in another. But separate sections for results and discussion are likely to lead to duplication or thumbing back and forth, because it is generally impossible to discuss the results without restating them. The results can usually be presented right along with the discussion of them. This arrangement has been used in the specimen report in Sec. 21.2.17.

When lengthy tables of numerical results must be included, they can often be put in the appendix (see Sec. 7.4). Then small tables may be excerpted from them, or significant or average values may be taken from them, and incorporated into the discussion. Similarly, the most important curves or graphs may be put into the discussion, while the more routine ones are relegated to the appendix (see Secs. 18.2 and 19.2).

The discussion that goes along with the results explains them and elucidates them. It points out any qualifications or limitations they have; it brings to light suspected sources of error; and it recognizes unexpected results and tries to account for them.

The discussion evaluates the results, and it interprets them and investigates their significance. In this process it probably arrives at conclusions—decisions and judgments based on the evidence presented in the report. If he has been authorized to do so, the report writer may make recommendations in line with his conclusions (see Sec. 2.12).

The conclusions that stem from the discussion should usually be stated in their logical place—at the several points where the discussion has led up to them. They may also be gathered together and restated (perhaps in a section labeled *Conclusions* or *Conclusions and Recommendations*) either at the end of the discussion, as in the specimen report of Sec. 21.2.15, or near the beginning of the report, in the summary (see Sec. 7.3).

A discussion seems more complete and is more satisfying to the reader if it comes to a logical and obvious ending on the level of the overall discussion rather than simply stopping on a note of detail. One kind of ending that is often appropriate is a *summary,* perhaps with emphasis on important results and conclusions. This summary stems from what has gone before; it is an integral part of the discussion, and it may depend on the discussion. It should not be confused with the *independent* summary, discussed in Sec. 7.3, which goes at the beginning of most formal reports.

7.3 Summary or Abstract

Most present-day reports of any length contain a synopsis very near the beginning, either just before or just after the foreword. This synopsis is sometimes called a *summary,* sometimes an *abstract*. Although these two terms are often used interchangeably, some people do make a distinction between them:

(1) A *summary* is a nonlinear reduction—a gleaning of the most important points, which may be reproduced full-scale from the body of the report. A summary is likely to be longer than an abstract and is usually found in an investigative or survey report where the reader is particularly interested in the results and conclusions.

(2) An *abstract* is a more or less linear reduction of the whole report. It is likely to appear in a research report where the reader is interested not only in the results and conclusions but also in the introductory parts and the procedure.

Of these two sorts of synopses, both of which appear in the specimen report of Sec. 21.2.16, the *summary* is usually more useful, and it is likely to be more suitable for formal reports. Therefore most of the discussion that follows will talk specifically about the summary. But in general the principles given apply to every synopsis, whether it is labeled *Summary* or *Abstract*.

7.3.1 Summary

The initial summary is a synopsis of the body of the report, with emphasis on important findings, conclusions, and recommendations. Therefore it should be written after the main body has been completed. *This summary is not to be confused with the summarizing conclusion that may form a part of the body; it is separate and distinct.*

The purpose of the initial summary is to permit the busy executive (or the busy underling) to get the significant information of the report at once, without having to hunt for it and without having to wade through the whole report. *Therefore the summary must be self-sufficient*—it must be independent of the rest of the report, with two possible exceptions: (1) If the foreword precedes the summary, as it usually does, the summary may depend on the foreword for background information, as in the specimen report of Sec. 21.2.17. Sometimes, though, the summary or abstract is put before the foreword, and made completely self-sufficient, as in the specimen report of Sec. 21.2.16. (2) The summary may state unsupported conclusions, the evidence to support them appearing later, in the body.

With these two possible exceptions, then, the summary must supply brief statements of: (1) the nature and purpose of the investigation (unless this information has been supplied in the immediately preceding foreword); (2) the facts or results; and (3) the chief conclusions and recommendations, if any (see Sec. 2.12).

Thus the conclusions and recommendations may be stated in a subsection of the summary, or they may have the status of a separate main section, taking the place of the summary. You will sometimes find statements of fact as well as decisions and judgments under the heading *Conclusions.* When decisions and judgments are included, they are probably the most important part of the report. Furthermore, they are inherently different from facts. Therefore the heading *Conclusions* should usually be reserved for the decisions and judgments, and the facts stated in a separate section of the summary. If this practice is followed, the conclusions and recommendations will form a suggested program of action.

The conclusions and recommendations may be set up as two separate subsections, or they may be put together. Since each recommendation is likely to follow logically from a conclusion or group of conclusions, we believe it is usually better to combine the two.

However the summary is arranged, it should contain only really significant material. It should not contain lengthy quotations of numerical data; if the report is largely numerical, the summary should contain only average figures or the most important figures, which may be arranged in short tables.

Since the body should itself tell a logically complete story, it must be self-sufficient and independent of the summary, too. Thus no information should be contained in the summary that does not also appear in the body.

Because body and summary are independent and each self-sufficient, there is usually repetition from body to summary. In fact, whole sentences from the body are often copied in the summary. Don't let this repetition worry you. Body and summary must each be complete, and many readers will skip one or the other.

Every summary should be as short as it can be without omitting essential information. The length of the summary will vary according to the length of the report and the amount of information in it. Thus no rigid limit can be set; but a summary should seldom exceed two or three pages unless the report is very long.

7.3.2 Abstract

The kind of synopsis that we have been calling an *abstract* appears in many formal reports, and it almost always accompanies the technical paper (see Chapter 12). The abstract often appears entirely separate from its original report or article in one of the journals such as *Chemical Abstracts, Science Abstracts, Biological Abstracts,* and *Horticultural Abstracts* that print nothing but abstracts from other publications. These journals have their own staffs of abstracters; but there is only one way to be sure that their abstract of your report or paper will say what you would like it to say: provide a suitable abstract yourself, along with the original publication. Thus the abstract, perhaps even more than the summary, should be capable of standing on its own feet.

Most abstracts seem to fall into one of two overlapping categories: (1) the *descriptive* and (2) the *informative*. The descriptive abstract describes the report or paper and tells what it is about. It is a sort of prose table of contents. The informative abstract, on the other hand, transmits some of the information—at least the most important points—of the original report or paper. The informative abstract is more valuable than the descriptive, and it is usually to be preferred.

Because the abstract is usually a more or less linear reduction, and because it is usually quite short, the reduction factor is likely to be large, particularly if the report or article is a long one. The greater the reduction factor, the harder it is to avoid the purely descriptive abstract. But you should try to get as much actual information into your abstracts as you can without making them excessively long.

As an illustration of the greater value of the informative abstract, here are two examples each of descriptive and informative abstracts.

Descriptive Abstract

A simple method of building close-packed molecular and crystal models is described. It has proved its pedagogic value as well as its many advantages to

Informative Abstract

Close-packed molecular and crystal models can be built easily and inexpensively by dipping balls of various materials into latex of 60-percent concentra-

Descriptive Abstract

the research worker. Two diagrams and four photographs are included. The method was evolved in the author's laboratory.

Informative Abstract

tion. Desired colors can be obtained by dipping the balls into a quick-drying water-insoluble paint before the latex is applied. Slight pressure is all that is needed to stick the balls together to form a model. They can be separated by a quick jerk.

Descriptive Abstract

Pulse hysteresis loops for a representative group of ferrites are shown and graphs presented for determining the value of pulse permeability to be used in pulse transformer design.

The assembly and circuit test of the ferrite-core pulse transformers are discussed and comparisons are made with Hypersil-core transformers.

Informative Abstract

Effective pulse permeabilities of those ferrites range up to a maximum of 450 for 0.1-microsecond pulses, while that observed for 1-mil Hypersil (with air gap) is approximately 100.

The assembly and circuit test of 1-to-1 and 3-to-1 ferrite transformers are discussed. The transformers described are equivalent in performance to Hypersil transformers, and yet are more simply constructed and more compact, and can be produced at considerably lower cost. Ferrite-core pulse transformers are admirably suited for 0.1-microsecond pulse applications.

7.4 Appendix

The appendix is a highly useful and important part of the report, even though it is shoved into the background behind the main body. The appendix is the place to put any material that needs to be included in the report but that is not an essential, integral part of the main presentation.

We have stressed the points (1) that every technical communication should be as simple, as short, as unencumbered as it can be and (2) that important ideas should not be buried under a mass of detail. The appendix makes it possible to unload detail information or information of secondary importance from the main presentation, yet still to include it for record purposes or for the sake of those readers who may want it or need it.

Because it plays this role, the appendix is sometimes treated as a dumping ground, and simply thrown together helter-skelter and stuck into the back of the report. It does the reader no good to stumble upon some miscellaneous additional information after he has finished reading the report. He should be given the opportunity of looking up pertinent information while he is reading the related part of the main body. *Thus every section of the appendix should be keyed to the text by a specific reference. Any material that is not connected to the text closely enough to be worth mentioning is not worth putting into the report at all.*

The sections of the appendix should be arranged in some rational order, often the order in which they are referred to in the text, and they should be numbered (or lettered) serially.

The appendix may contain material such as:

(1) Tables, graphs, or illustrations that are an integral part of the main story but which will not conveniently fit into the body of the report.

(2) Tables, graphs, or illustrations that are not used in the main story but which are of interest because they form the basis of it or are related to it.

(3) Tabulations of data that are presented in graphs in the main body.

(4) Detailed descriptions of equipment or procedure, when the body contains only a general or overall description.

(5) Descriptions of rejected methods.

(6) Samples of forms, data sheets, or questionnaires used in the investigation.

(7) Derivations of equations.

(8) Sample calculations.

(9) Tables of symbols or definitions.

(10) Copies of speeches, contracts, exhibits, or company literature.

(11) Bibliographies.

(12) Any material that must be included for record purposes.

7.5 Foreword or Preface; Letter of Transmittal

The terms *foreword* and *preface,* as applied to reports, are used interchangeably. Since *foreword* is the more common of the two, it will be employed in this discussion.

Most formal reports contain either a foreword or a letter of transmittal, sometimes both. When the letter of transmittal alone is used, it performs the functions of the foreword (Sec. 7.5.1) as well as its own special ones (Sec. 7.5.2).

7.5.1 Foreword

You will notice that the functions listed for the foreword overlap those listed for the introduction phase of the body. The foreword and the body of each report should be planned together so that they complement each other. There should not be any substantial amount of duplication from one to the other. (Contrast this relationship with that between body and summary: the summary covers the same ground as the body, and may be lifted from it word for word.)

The foreword does some or all of these things:

(1) Provides the reader with general orientation, as contrasted with the complete background information contained in the body. *It tells him broadly what the report is about;* that is, it expands on the title and probably contains a brief statement of basic purpose.

(2) Tells who authorized or requested the investigation to be made or the report to be written (if such information is relevant).

(3) Makes acknowledgments for help received by the author (some organizations use a separate section for acknowledgments).

(4) Makes any pertinent comment about the report.

(5) Refers to other related reports, either those already in existence or future reports that are going to be written about further developments of the same investigation.

7.5.2 Letter of Transmittal

The letter of transmittal is often used in a more specific and intimate sense than is the foreword. Because it is a letter, it is addressed to a specific person or group and exhibits the characteristics of a business letter. A solicited report—that is, one that has been assigned to the writer—frequently is accompanied by a letter of transmittal.

Composition of the letter of transmittal varies widely among those organizations that use this form rather than a foreword. Besides performing the functions listed above for the foreword, the letter of transmittal has the specific job of *transmitting* the report; that is, it says "Here is a copy of the report . . ." or words to that effect.

In addition, the letter of transmittal often refers to specific parts of the report, and it may repeat particularly important points such as conclusions and recommendations. It may also repeat such introductory information as statements of purpose, scope, and limitations. However, the letter of transmittal should not be depended on to transmit any essential information not stated in the report proper, because some readers are likely to skim the letter or to skip it entirely.

When the letter of transmittal contains any appreciable amount of information—that is, when it is in any sense a part of the report—it is usually bound into the report, often immediately after the title page. Sometimes, of course, a report that is complete with a foreword is mailed along with a letter that says simply "We are sending you the report. . . ." This kind of covering letter, too, might be considered a letter of transmittal, but it is usually not bound into the report.

A specimen letter of transmittal is reproduced in Sec. 21.2.18.

7.6 Table of Contents

The table of contents, which should be included in every report of more than three or four pages, does something more important than simply locate page numbers: it provides the reader with an outline, or an overall plan, of the report. Just as you can do better justice to a banquet if you have been supplied with a menu, so you can read a report more intelligently and more efficiently if you know at the beginning what to expect and where to expect it. The table of contents furnishes the reader with the framework to which he can fit the parts of the report. Because it performs this important function, it should be constructed carefully.

The table of contents is compiled from the headings in the report. Its entries should be duplicates of these headings, including any number or letter symbols, and they should be arranged in the same outline plan. Thus the contents should clearly indicate the relative weights of the various headings and subheadings, and show their relationships with each other.

No entry should appear in the table of contents that does not also appear as a heading in the text. If it does, the reader may be frustrated and annoyed when he turns from the contents to the text and hunts for a heading that is not there. On the other hand, it is not obligatory to enter every text heading in the contents. Although none of the major headings or subheadings should be omitted, it may sometimes be appropriate to leave out the lightest-weight subheadings, especially if they seem to obscure the main outline; but the headings of any one degree should either be all listed or all omitted.

The table of contents of this book furnishes an example of a table of decimal-system headings. For a table of headings differentiated by typography and indentation, see the specimen report in Sec. 21.2.17.

The table of contents is not to be confused with the *index,* which lists its entries in alphabetical order, and which usually appears at the end of the book rather than the beginning. Only very long or complex reports require indexes.

7.7 Cover; Title Page

The formal report customarily has a binding or cover, usually of cardboard, and a title page, both of which carry identifying information. This information consists logically of several blocks, and it can be clarified by being arranged in physical blocks according to its logical divisions. Furthermore, the more important pieces of information—particularly the title itself—should be emphasized by larger type size or by capitalization. These principles are illustrated in the specimen reports of Secs. 21.2.15, 21.2.16, and 21.2.17.

Secs. 7.7.1 and 7.7.2 list the *minimum* requirements of identifying information for cover and title page, and suggest a few possible additional items. For these pages to do an effective job, they should be as uncluttered as possible so that the important information will stand out boldly and clearly. Therefore all nonessential items should be removed from the cover and title page and either omitted or put somewhere else in the report.

7.7.1 Cover

The cover should have a label that provides enough information so that the people who have to handle or file the report can identify it without looking inside. Minimum requirements usually are:

(1) Title
(2) Name (and perhaps address) of issuing company or institution
(3) Date of issue
(4) Report number (if any)

In addition, some organizations number the copies of each report serially and put the copy number on the cover.

The covers of students' reports must also carry:

(1) Author's name
(2) Name and number of course or subject

7.7.2 Title Page

The title page repeats all of the items on the cover. It may also carry additional pertinent items such as:

(1) Subtitle or elaboration of main title
(2) Author's name
(3) Name of man or company for whom report was prepared
(4) Further details about issuing agency
(5) Contract or project numbers
(6) Any other necessary information

But remember, do not let the title page get cluttered with unnecessary material.

7.8 Other Elements of the Report

The term *element* is used here to mean any part of the report that has a specific function and therefore needs the writer's particular attention.

It should have been made clear by now that the material being used determines largely the form that the main part of the report will take. The report form should be flexible and not something into which material must be crammed. But the report writer new to the job may not know exactly what to look for. Other elements besides those already discussed should be considered to see if they should be used.

(1) In addition to describing the object, scope, and general purpose of the report, you may need to include a sizeable section on *importance*. Reports are frequently used in connection with technical sales literature; one group of readers may need to know how important you consider the material to be. Some companies, in fact, require that their reports always contain such a section so that the information under discussion can be evaluated and its worth more clearly understood.

(2) When a project cannot be carried out without the help of specially trained people—technicians, specialists, supervisors—you should include in your report a section on *personnel*.

(3) Because a report is written for practical purposes and frequently read by executives who have finances in mind, don't neglect the possibility of

including a section on *costs* and *expenditures*. Sometimes recommendations alone do not tell the entire story. If an estimate of costs will have any effect on the overall report, such a section should be included.

(4) We have stated that the formal report frequently contains a section on background in order to link the present problem with what has been done in the past. Many company reports call for a statement on the future as well, labeled *Forecast*. This can be in the undergraduate report, too. It will follow the evaluation of results, conclusions, and recommendations, and will be an extension of them. The forecast is especially needed in the research report; here you can carry your recommendations still further by discussing probable trends and what may be accomplished in the future.

7.9 Outlines of Company Reports

Although form in the formal report should always be subordinated to material, it is a good thing for both student and company writer to know how some companies have designed their reports about their specific products. *The Hercules Technical Reports,* a guide to the preparation of reports put out by the Hercules Company, in its preface cites four steps that seem essential in carrying out a competent investigation:

(1) Understand at the outset the objective of the investigation.
(2) Determine the key technical problems to be solved.
(3) Direct the experimental program to solve these problems.
(4) Study the results and draw the correct conclusions from them.

At another place this guide states that "all of your technical reports should have one chief objective—to provide the technical facts and recommendations that must be used by the company management, along with economic and other factors outside the scope of the technical investigation, in deciding on a course of action on the problem you have studied."

The outline given below has been developed as a general pattern for all Hercules technical reports:

Introduction
What is the *objective* of the work?
Why is it *important* to the company?
What is the *background?*

Discussion of Results
Briefly, what work was done?
What were the most important results?

Conclusions and Recommendations
What do the results mean?
What action do they suggest?
What future work is planned?

Experimental Section
How was the work done?
What results were obtained?

Bibliography
What pertinent literature was consulted?

Again it must be stressed that the above outline has worked out best for this particular company, as the following has for the Chrysler Corporation:

Sections and Headings	*Function*
Introductory Section Title Page Preface Table of Contents	Gives background information; lets reader know what the report is about.
Summation Section Subject Object Conclusions Recommendations	The heart of the report; gives an overall view of • what was studied. • what the purpose of the study was. • what you found out. • what you think should be done.
Technical Section Parts or Materials Tested Apparatus Procedure Results Discussion	Give details about • what you tested. • equipment you used. • how you did the work. • what happened. • what was your interpretation of the work.
Graphic-aids Section Tables of Data Curves Drawings Photographs	Gives visual aids in communicating efficiently.
Supplementary Section (Appendix) References List of Symbols Sample Calculations	Presents useful information that is not an integral part of the report.

7.10 Check List for Reports

In our own work in preparing reports, we have found a check list invaluable. We pass it on to you. After you have written a first draft, see how well it will stand against the suggestions we are now giving you.

(1) *Summary*. Make the summary fit your piece of writing. Go through your data, pick out the salient facts, and especially the conclusions and recommendations. Include them in the summary.

(2) *Introduction*. Relate the introduction to the needs of the reader. Try to imagine what he will want to know first. Show the purpose and scope. Identify the subject matter. Relate it to other projects. Show the basic methods and procedure followed in carrying out the project.

(3) *Headings and Titles*. Break the report into parts by using informative headings. Then organize the material into logical units. They make reading easier by providing white space. They make it easier for the reader to refer back to particular sections.

(4) *Paragraphing*. If the main idea in a paragraph can be expressed in a single, highly compressed sentence, the paragraph has achieved unity. Pay attention to topic sentences.

(5) *Listings and Tabulations*. Don't ask the reader to dig out details. Assist him by numbering various steps, by vertical listing, underlining, italics, capitalization, spacing, and similar devices.

(6) *Point of View*. As much as possible, adopt one style and one attitude toward both your material and your reader. Don't shift from an objective point of view to a personal one, or from a negative to a positive attitude, especially within a short space.

(7) *Transitional Material*. Supply phrases and guide words to lead the reader from one thought to another. Provide headings, listings, and tabulations. Don't be so sparing and limited in guides and signs that you are practically unintelligible.

(8) *Figures and Tables*. Determine whether tables and figures should be in the text itself or should be reserved for the appendix. Improper positioning within the text can create confusion for the reader.

(9) *References and Bibliographies*. Be consistent in the method you use to record references. When you are working on a report over a long period, you can make mistakes in the bibliographic system. Pick an appropriate system and remain faithful to it in any given report.

(10) *Reader Appeal*. Check the overall effectiveness of your report:

(a) Have you devised a method of getting the reader to start reading? This question points up the importance of a good introduction.

(b) Have you devised methods to keep the reader reading? Are your organization and structure as clear and logical as possible?

(c) Have you decided what you want your reader to retain? This question will be a guide to how you should emphasize material in abstracts and summaries, appendixes, tables, and graphs.

EXERCISES

1. You may have to write a long formal report for your course in technical reporting. Choosing a suitable subject is sometimes difficult. Write a letter to your instructor telling him of your academic background, your professional ambitions, and any special interests you have. Such a letter will help him to advise you on a subject for your formal report.

2. Your formal report should evolve from as real a situation as possible and should be directed to a real company, group, or individual who can use your information. Set up such a situation.

Assume that a person in authority asks you to supply information that he or his company can use. Write yourself a letter assuming that it is coming from such a person. The letter should explain why you have been asked to do the project and write the report, how much ground you are to cover, what the information is to be used for, and who will be reading the report.

Such a letter will give you proper motivation in writing your formal report.

3. Answer the letter you have received, accepting the assignment, and include a statement of what you are prepared to do.

4. Write a memo report to your instructor. Tell him about the report you have been asked to prepare. Give him as complete an analysis as you can at this time of the material you will be using, the sources of information available, and a tentative outline of the report.

5. Read a report in your field. Write a memo report to your instructor about your reading. Discuss your reading in respect to what has already been said in this book about formal reports. Include with your memo report an informative abstract of what you have read.

6. At several points during the course, write progress reports on your formal report to (a) your instructor and (b) the person or company making the assignment.

8. Laboratory Reports

First let us distinguish between the laboratory *report* and the laboratory *notebook* or *log*. The notebook is a complete record for yourself (or your company) that is filed away and practically never given out for others to read. The report is a vehicle for transmitting to other people the significant information of the notebook. It may be a formal report or an informal report, a letter or a memo.

This chapter is primarily about the college laboratory report. It also contains a brief section on the notebook.

8.1 The Notebook

Your laboratory notebook should be as much as possible like the notebooks used in actual research laboratories. It should contain a running record, or diary, of the work you have done in the laboratory, with entries in chronological order. *This record should be so complete and so clear that it will enable you to duplicate the work after a lapse of years.*

For each project or experiment, the notebook should probably contain these four elements:

(1) A *very brief* statement of the fundamental nature and purpose of the investigation—what it's all about.

(2) A clear but succinct record of what you did (and when you did it), including such details as instrument numbers, diagrams of your setups, and calculations.

(3) A record of your results. Numerical data should be supplemented by curves, which may often advantageously be plotted during the conduct of the experimental work.

(4) A *brief* discussion of unusual or particularly noteworthy results or observations.

Unless you are instructed to include them, the laboratory notebook should *not* contain:

(1) Lengthy presentations of theory or textbook material.

(2) Extended discussions of ordinary or routine results.

Like all other forms of technical reports, the laboratory notebook is clarified by a liberal use of headings, sketches, and diagrams.

8.2 The Report

The reports required of students in college laboratory courses may be designed primarily for two different purposes: (1) to serve as a realistic exercise in report writing—to prepare the student for writing the reports that will be demanded of him later, when he is working in industry or in a research laboratory; (2) to serve as a further tool for the teaching of the subject of the course—to drill the student in the details of what he is studying in the classroom.

These two purposes can be combined, and a realistic report-writing exercise can at the same time help to teach the main content of the course. If the assignments for reports are planned with this end in view, the reports will be real technical reports. They will transmit information, and they will have all extraneous material weeded out. They will follow the principles set forth in this book.

Unfortunately, however, the report assignments in many laboratory courses have been laid out primarily as teaching aids, with little thought to their value as report-writing exercises. The resulting reports are likely to be quite unrealistic. The rest of this chapter is addressed primarily to students (and instructors) in those colleges that have not set up realistic reporting procedures in their laboratory courses. If you are one of these students, you will of course write reports that follow the instructions you receive in class. But it is important for you to know how such reports depart from good technical-reporting practice, because all too many engineers and scientists carry over from their college laboratory experience a mistaken idea of what a technical report should be.

(1) The "teaching-aid" laboratory report usually contains a lot of extraneous material, and its elements are in a poor sequence. Here is an outline typical of this kind of report:

(a) Purpose or object of test
(b) Discussion of basic theory
(c) Complete identification of all machines and instruments used
(d) Detailed description of methods used
(e) All data in tabulated form
(f) Results in tabulated form
(g) Plots
(h) Conclusions and discussion
(i) Appendix: computations

A report that follows this outline is largely a transcript of the laboratory notebook. Notice that the most important items—the conclusions and discussion—are badly buried under a lot of detail information. Many of these details might better be omitted from the report entirely, remaining on record in the notebook; or at least they should be relegated to the appendix. In particular, the names and numbers of machines and instruments are seldom significant, and neither is a *detailed* description of method unless the method is new or unconventional. There is seldom any need for a complete tabulation of all data taken, particularly if they are clearly shown in plots. Usually averages, or final results, are sufficient. And if an outline like this is followed rigidly, as it is in some colleges, it gives the student the erroneous idea that the preparation of a technical report consists of little more than filling in numbers in a ready-made form.

Because it is supposed to drill the student in his classroom work, the teaching-aid report is likely to contain a long discussion of basic theory—usually just a rehash of material that can be found in any textbook on the subject. The real technical report, on the other hand, presents only as much basic theory as its specific audience needs. Because the teaching-aid report is designed to demonstrate to the instructor that the student has read the assignment and done all the required work, it is likely to contain all the information the student can think of on the subject, important or unimportant. Indeed, under these circumstances the student would be foolish to do the pruning job essential to a good technical report.

(2) The audience of the teaching-aid report is an unrealistic one. The author of a real report generally knows more about his subject than his readers do. In fact, he may even have to explain his vocabulary to them. But the student writes the teaching-aid laboratory report to his instructor. Thus the audience probably knows more about the subject—at least in general terms—than the author. The author certainly does not have to furnish general background information or to define terms.

Students can be instructed to make their laboratory reports as much as possible like the reports issued by industrial and research organizations: (1) the laboratory report can follow a flexible outline like the one suggested in Chapter 7; (2) it can be made as brief as possible without omitting essential information; (3) it can be addressed to a specific hypothetical audience—say to the other members of the class, who may be presumed to have the same general training as the writer but to lack his knowledge of the immediate subject. Such a program requires careful planning by the instructors in charge of the laboratory—planning not only of the reporting assignments but perhaps also of the experiments themselves. The effort involved is well worthwhile. But remember, we are not advising you—the student—to disobey the rules of your own laboratories.

9. The Thesis

At one time or another you will probably have to write a thesis. Undergraduate courses often require term papers that are theses in capsule form. Many colleges, especially engineering colleges, require a thesis for graduation. If you are working toward an advanced degree, you will certainly find the thesis a major requirement. And many persons in industry are taking extension courses and advanced degrees requiring theses.

There are strong resemblances between the thesis and the report, and the two are often completely alike. Strictly speaking, a thesis is a document containing information that will advance human knowledge, regardless of when that information will be used. A report, of course, contains highly practical information for more or less immediate use.

Just as the report has become formalized, so has the thesis. But not all thesis forms are alike. Be sure to find out what your college demands. Most colleges publish manuals giving specific directions to fit their individual needs. Before we go into a discussion of thesis writing, therefore, we would like to make these points:

(1) Many of the principles cited in connection with the report apply equally well to the thesis.

(2) Everything said in Part III on *Tools and Methods* applies to the thesis.

The complete job of writing a thesis will include the following phases: selecting the subject, consulting the adviser, gathering the material, arranging the material, and actually writing the paper itself.

9.1 Selecting the Subject

Selecting the subject takes a lot of time and care. If you are an undergraduate, you may feel that thesis writing is a bothersome chore, devised by teachers as one more task before you graduate. But if you are doing postgraduate work, you will by now have the incentive to write a good thesis; and if you are employed, you will have the experience and the insight to see that through the thesis you can make a real contribution.

But you may not be able to select your thesis subject immediately. You will have to reject many ideas before you come upon one that fits the situation. Here are some principles that should guide you:

(1) *The project should require some real research on your part.* Probably the two largest sources of research are experimentation and the study of documents. If you already have a subject in mind, see if it is of such a nature that you can do something in the way of original experimentation in the laboratory. It is doubtful whether a series of laboratory reports will fit your needs. Many laboratory experiments are routine; they have been devised largely as practice work. Your thesis requires that you set up apparatus, use materials, and devise a working procedure that you, not someone else, originated.

These comments apply to library research as well. A paper summarizing the contents of a book is not a thesis; if you find all your material in one or two articles in a reference book, you are performing little genuine research. You must weigh and judge and evaluate your material to see if it fits this first criterion of being a *research* project.

(2) *The research project should be interesting and beneficial to you.* We have indicated that undergraduates sometimes have difficulty selecting good research subjects. They go about the job in the wrong way. A speaker usually makes a poor speech if he discusses a topic that holds little attraction for him. So it is with you as a writer. If you have to plow through your material, fighting boredom all the way, undoubtedly you will not come out with anything very satisfactory.

The subject you pick and the work you do should be useful to you. It should add to your store of knowledge; it should stir your imagination; it should enrich your class work or your professional life.

(3) *The research project should be worthwhile and of real value.* In addition to looking at your thesis from your own point of view, try to get an objective evaluation. Do you feel that the experiments you perform will fill a gap in the field under consideration? Would there be any point in another person going through the same process? Is it likely that anyone in the future would be inclined to begin another project where you left off?

Especially, is it possible to approach your subject in a scientific manner? Research implies a scientific attitude, open-mindedness, and a readiness to accept changes as they occur. If you start a project with a firm conviction of what you will find, you will probably end up with something that is neither good research nor good writing.

(4) *The subject should be commensurate with the limitations of time, length, cost, and available data.* Pick a subject that can be handled within an allotted time and space. If your thesis (perhaps it is actually a term paper) is required for only one of four courses, you certainly would not be expected to provide as much material as if you were writing for a master's degree. If it is to be completed in one term, don't choose a topic so long that it can be covered only in two terms.

Thesis material is found in many places, and you should carry out an exhaustive search to find it. But there are limitations. Is the data available? Can you, for example, find the material in your college or company library, or must you use outside agencies to help you?

Cost is sometimes a deciding factor. Materials and apparatus may be costly; books may have to be bought and material reproduced. Think of all these contributing factors in choosing a thesis topic.

(5) *Your thesis subject should fit local requirements.* The thesis writer is in a peculiar situation. Not only must he write for the remote reader, he must write under the supervision of an adviser who will determine whether the material is acceptable. And each adviser will have established over a period of time a set of standards by which he will judge the work submitted to him.

(6) *The thesis writer must know where to look for suitable topics.* One source of acceptable undergraduate thesis material is any job experience you have had, or perhaps a part-time job you are doing. Such practical experience opens up many interesting and profitable topics.

Think of trends in research. Think of the many things still to be done—machines to be invented, processes to be perfected, scientific horizons to be broadened.

Another fruitful source of suggestions is the person who is an authority in a specific field. Your teachers and supervisors can tell you about neglected areas of information. Writers often make provocative statements or show up contradictions and inconsistencies that should arouse your curiosity. Think of these in terms of possible thesis subjects.

9.2 The Role of the Adviser

Don't look upon your thesis adviser as a person who is to do your work. He is a counselor, not a coauthor. He has a knowledge of what has been done in your field; thus he saves you time and also introduces you to new ideas. He sees you in relation to other thesis writers; in this way he prevents repetition and duplication. He knows the requirements of your college; he will show you how to arrange your material and how to put it into acceptable form. Above all, your adviser has written a thesis or two himself and can point out the pitfalls.

But treat him as an adviser, not as a proofreader. Listen to what he has to say. When you need information, ask for it only if it cannot be found elsewhere. And, especially, don't expect him to revise your manuscript for you. Nothing will irritate a thesis adviser more than to be mentioned in the acknowledgments as the person who corrected your grammar and punctuation. He should not take the responsibility for a writing job that is yours, and yours alone.

9.3 Gathering Material

The next definite phase is the gathering of material. This is a period when facts and ideas, results and findings, come in from all sides—from books you read, tests you run, the work of other people that you look at.

At the beginning of this phase, look at your problem and try to estimate how big a job it will be. Get oriented to it. Ask yourself some questions. How much laboratory research does the thesis involve? Often there is the problem of repetition. Should you rerun some of the tests that other people have done, or can you assume that they are definitive and correct?

The question of field trips comes up, too. If you are a conscientious researcher, you should visit factories and plants to see practical applications of your ideas. You will have to decide how much of this activity is advisable within your limits of time and money.

9.3.1 Interviews

Will interviews be useful to your project; and, if so, how should you handle them? When you are interviewing a busy man, remember that he is granting you the favor. First of all, make the interview easy for him and for yourself by preparing for it. When you get in touch with the person to be interviewed, make it very clear why you must talk with him personally. Prepare a set of questions—not too many, ten or twelve perhaps—and break them up in such a way that they can be answered directly and decisively. Don't trust your memory. Write the answers in a notebook; your man will appreciate having his remarks recorded accurately. And be particularly careful in transcribing statistics and direct quotations.

9.3.2 Questionnaires

You may find it necessary to use a questionnaire as a source of information. Like the interview, the questionnaire must have a reason for being used. You must convince those to whom the questionnaire is directed that you need the material and will be grateful for getting it. Write a covering letter incorporating these two points. The questions should be phrased in the form of a list, not in lengthy paragraphs. There should be comparatively few questions, they should be as brief as possible, and above all they should not be "loaded." A question that practically tells the reader how to answer will be of very little value to you.

Advertising writers put a great deal of effort into the form their questionnaires take. They number the questions, phrase them in parallel form, and provide boxes for positive, negative, and undecided answers. If you feel that you need to know more about the technique of preparing the questionnaire and evaluating it, consult the appropriate books in your library.

9.3.3 Library Surveys

While you are in the library and before you get involved in your project, look over library research possibilities. We have pointed out that your choice of subject may depend on how much help your library can give you. Find out if the library can also help you obtain books from outside sources, such as a state library or special foundation libraries. Find out too if any material you need is restricted or confidential.

You now come to a second step in the gathering of material: a thorough survey of existing literature connected with your subject. The survey will not only be valuable for direct information; you will probably be required to discuss it as a part of your thesis.

Remember that you are building on a base of research and experimentation that has been done by other people before. Your readers will want to know who these people are and what they have accomplished. Track down all available books, articles, reports, and pamphlets. First of all, consult the card catalogues. Then refer to the periodical indexes, bound volumes containing brief synopses of articles published over specified periods. The articles will be cross-referenced like the items in the card catalogue. Two of the better-known periodical indexes are *The Reader's Guide to Periodical Literature* for articles of a general, nontechnical nature, and *The Index for Applied Science and Technology* for scientific and engineering articles. There are also reference works on more specific subjects, such as *Chemical Abstracts, Biological Abstracts, Nuclear Science Abstracts, Psychological Abstracts, Meteorological Abstracts and Bibliography,* and *Scientific* and *Technical Aerospace Reports* (Abstracts). This is not an all-inclusive list; the items are representative of references available in all fields in engineering and science.

Other indexes available are: *International Index to Periodical Literature in the Social Sciences and the Humanities;* the *British Technological Index* for articles in British technical journals; and *Computing Reviews,* bibliography and subject index of current computer literature.

Indexes such as these will provide information on what has been done before and on material that you may be able to use in a new way. These sources may supply the bulk of your material or may be used to substantiate experimental work you are carrying on.

You will, of course, place your information on note cards, as we have suggested in Sec. 3.1.

9.4 Arranging Thesis Material

The main portion of the thesis—the body—resembles the body of a long report. But remember that the motivation for writing it is different, and it is the motivation that will determine the way in which thesis material is arranged.

9.4.1 Introduction

In this section write a statement about the project, defining it, limiting it, and establishing how important and valuable it seems to you. Just as in a report (see Sec. 6.1), write an early statement of important results or conclusions.

9.4.2 Survey of Literature and Recorded Data

This section should mention all the references and documents having any bearing on your specific project. Here the reader will find a comprehensive survey of what has been done by others: what they have written and said, graphic material they have included, and all the statistical detail of past projects.

But you may be asked to do more than mention these records. You may be expected to give an objective interpretation of them. A distinction must be made between bibliography and the survey of literature. The bibliography, placed at the end of your thesis, is a list of the documents that assisted you specifically. The survey, on the other hand, is much larger. It contains not only your specific bibliography but also a record of everything that has been done in the broad field of which your thesis is only a part. Citations must be made by person, date, and place (see Secs. 17.5 and 17.6).

9.4.3 Report on Research

This is the largest part of the thesis. First of all, discuss any special characteristics of the thesis research that have not been treated in the introduction. For example, it may be necessary to mention here that the project was carried out under special temperature conditions or that a new laboratory technique had to be devised.

Then cite any theories, hypotheses, or assumptions. In experimental work, particularly, the reader must know the theoretical basis upon which the entire project rests.

Frequently experimental work requires special laboratory apparatus. If you designed new testing apparatus, this is an important point that must be described and discussed.

After the discussion of theories and materials, describe the procedure you followed, whether your project is library research or experimentation. Follow the best practice of technical description by first giving the overall view, and then taking up each step or phase in more detail.

Because the thesis at least presumably describes an original investigation that explores new territory, it is likely to involve new or unconventional methods. Therefore the description of procedures followed often occupies a more prominent and a larger space in the thesis than it does in the usual technical report, which is more likely to involve standard or familiar procedures (see Sec. 7.2.3).

In experimental work, it may be necessary to run tests of one kind or another. These should be described, and the sequence clearly shown. Then give the results of the tests, putting the data as much as possible in charts, tables, and other graphic forms.

9.4.4 Discussion

Thesis advisers say that the weakest part of many theses is the discussion of results.

Make the discussion as strong and valuable as possible by being objective, although here is the one place in the thesis for presenting opinions and judgments. These must, however, be based on the results of experimentation and research and not on any wishful thinking. Show what your findings mean to the entire area of knowledge, and, if possible, discuss them in terms of the literature you surveyed at the beginning of the thesis.

In the discussion, too, you will evaluate the project as a whole. Don't be afraid to discuss any poor features you feel it has, any disappointments you encountered. State what you think may result from your findings. If you think that any practical applications may evolve, discuss them and suggest methods for extending your work. In that way, you will be helping the next thesis writer.

9.4.5 Summary

The thesis usually ends with a summary. Such a terminal summary will be a gleaning of the most important points taken from the body of the thesis, listed and tabulated for easy reading (see Sec. 7.3).

9.4.6 Other Elements

The typical thesis contains most of these supplementary elements: title page, usually containing a statement of submittal showing for whom the thesis was written; table of contents, often starting out with a thesis statement, which is a one- or two-sentence summary of the thesis subject; preface; abstract; bibliography; footnotes; index; and tables. These elements are discussed in Chapters 7, 17, and 18.

10. Instructions

10.1 Style and Form

Engineers and scientists frequently become involved in instructional writing, a kind of exposition that tells the reader to perform an action, or at least tells him how it can be done. Although technical writers are employed in many companies to put together manuals and instruction books, in other companies engineers must do it. In any case, engineers provide the raw material.

Even in a report, you may be called on to give instructions on how a laboratory test is to be carried out. In style and form, instructions range from the formal, terse set of commands so often found in business memoranda or posted in college laboratories to the informal how-to-do-it article printed in popular magazines.

In structure, instructions can be as simple and direct as "Turn out the lights," and as complex as the multipage manual on the maintenance of an airplane.

In any case, the first principle is to present the individual points as briefly as possible, and to make them clear and capable of easy reading. After all, you cannot afford to wade through an essay on fire prevention if the house is on fire.

A second principle is to write in such a way that the instructions are authoritative and the reader is easily persuaded to carry them out. Most people prefer to know why they are being asked to do things, and when they know they are usually convinced.

Instructional writing profits by most of the devices listed in Chapters 2 and 4. You may have to define your terms before you can issue orders. If a machine is involved, it is usually necessary to describe it and explain how it works. Processes, too, are frequently involved in instructional writing; study the techniques for explaining processes (see Sec. 4.2) and apply them to this kind of writing. In particular, remember the advice of Sec. 2.4 to deal with the whole before getting down to any of the parts.

10.2 Formal Instructions

Any instructions should be well planned in advance and put in a step-by-step arrangement. You will only confuse the reader and make the result unpredictable if you neglect to mention an important step at the right time.

Use simple language and aim it at the particular people who will be carrying out the project. It may be necessary at times not only to give positive instructions, but also to include explanations and precautions. Explanations will satisfy the reader's desire to know why he is doing a particular thing; precautions will save him from errors in judgment and sometimes from bodily harm.

Formal instructions form the basis of industrial manuals and instruction books. Therefore the elements suggested here will be useful to the student and to the industrial writer alike. Specimens of formal instructions are reproduced in Secs. 21.2.19 and 21.2.20.

10.2.1 Title

The title should be short and positive and should explain what the instructions are intended for.

10.2.2 Purpose and Scope

It is advisable to include a short introductory paragraph, explaining the purpose of the instructions, how much ground they cover, and what the result should be.

10.2.3 Equipment and Materials

It may sometimes be necessary to follow the introductory portion with a section of general information especially concerned with equipment and materials, particularly if these must be devised especially for the project.

10.2.4 Specific Instructions

Especially in simple problems, and at the same time for more technical material, specific instructions should be stated in the imperative mood—command language. For example, in giving instructions for making a sundial, do not say, "You should cut the design in two along the 12-o'clock line." Instead, say, "Cut the design along. . . ."

10.2.5 Explanations and Precautions

Explanations of apparatus and materials to be used in carrying out the instructions are usually placed in the indicative mood, the language of discussion as distinguished from the language of command. For example, at the end of his instructions about making the sundial, when he felt the need for further explanation, the writer changed from imperative language to say, "Through the year the sun and our clocks do not exactly agree. Sometimes the sun is 10 to 15 minutes ahead, sometimes behind."

On the other hand, if a precaution is of vital importance, don't hesitate to put it in the imperative mood—and underline it and type it in capitals, if necessary.

10.2.6 Numbering

Formal instructions are usually numbered, with each step placed in a separate paragraph.

10.2.7 Other Devices

Instructions within a specific section may be capitalized, underlined, or otherwise set off to distinguish them from the explanations and precautions. The latter, for example, are often indented beyond the usual margin, or put in separate paragraphs, or placed in parentheses. If not so important, they can be held over until the end.

Some formal instructions require check points. These sectional headings indicate where the reader is to check on what he has been doing up to a particular point. A drawing or some means of visual representation can help the reader decide whether he is on the right track.

10.2.8 Visual Presentation

Instructions are often clarified when accompanied by illustrations, charts, and other graphic material (see Chapter 19).

Of the two examples of formal instructions in Secs. 21.2.19 and 21.2.20, the first, written by a student, illustrates how even simple instructions can benefit from a formal style. The second—a more sophisticated example—is from an industrial manual.

10.3 Informal Instructions

The wider the reading public to whom instructions are addressed and the less compulsion put on the reader to follow them, the more you will resort to the techniques suggested in Chapter 12 for the technical article. You may have to use a vocabulary containing a minimum of technical terms and many words of high imaginative content. Often the first paragraph, with its general introduction to the subject, must be slanted to attract attention and convince the reader of the value of the project. All the devices of adaptation and popularization—analogies, examples, incidents—should be employed.

But when actual instructions are given, the formal method is customarily used.

10.4 The Simple Manual

Suppose that you were asked to write a simple manual. How would you go about it?

First of all, analyze your problem and decide to what extent your reader must be instructed. If a page will do, do not try to be fancy and write a half-dozen pages of elaborate directions.

Start writing the manual only after you have made a careful study of the process involved. For example, if you are writing instructions for a machine, operate it yourself, talk to others who have operated it, and in general profit by these experiences—both the successes and the mistakes—before you make up the format of your manual.

Gain as much information as you can from all sources. Investigate the experiences of others. See if a set of instructions already exists. Even if it is not satisfactory, it may give you a framework on which to hang a new edition. And don't neglect interviews, questionnaires, and other personal sources.

How-to-do-it writing (and manual writing falls into this category) benefits tremendously from graphic material. The quantity depends, of course, on how complicated are your instructions, how large a distribution you will have, how much money you can spend. But certainly drawings of machines, and pictures of people and devices in action, will help the reader to carry out the instructions. You must decide which parts of your write-up will be helped by illustrations and where you can obtain them. For a more detailed discussion of graphic material refer to Chapter 19.

As soon as possible make a rough outline of the manual as a whole. Decide on the approximate number of pages you will have; this will be determined by the complexity of the device, the time available, and costs. Investigate the various kinds of reproduction available; see Sec. 17.11.

Before writing the manual, experiment with a few tentative layouts and format designs. Decide whether to use the entire page width, or double, or even triple, columns. Decide on illustrations, and indicate where they will be placed. As in reports, headings are almost mandatory; show where they will appear. They may be only tentative, but they will show you how to arrange your material.

The next step involves preparation for writing, although you will probably have done some writing already. And here a clear outline can save you a lot of work. At the same time, adjustments are constantly made between the outline and the manuscript right up to the final draft.

Remember, too, that the three kinds of expository writing are usually found in a manual: *informative,* which is the basis of so much report writing; *instructional,* which we have been discussing; and *persuasive,* as used in articles and sales literature.

The first part of a manual, then, will probably be a combination of informative and persuasive writing. It will be an introduction orienting the reader to the purpose and function of the manual and the procedure with which it is dealing. The style of the introduction will depend on the knowledge the reader already possesses and how necessary the manual will be to him.

This section may be followed by one dealing with materials and equipment or descriptions of other items necessary to carrying out the instructions. Devices already recommended in this book should be examined—lists, tabulations, and so on.

The next step involves the detailed instructions. Pay particular care to their arrangement, and determine whether the techniques of formal instructions, or those of a more informal approach, should be followed.

In the actual writing, use the general principles already recommended. Use the imperative mood as much as possible. Assume that the reader knows little about the project itself; therefore be simple, and include analogies and comparisons. Make a careful check of your details to be sure they are accurate. Many procedures do not succeed because the directions are incomplete and inaccurate.

The final step in the preparation of the manual will be one of checking. Ask someone to go through the procedure as you have roughed it out. Only when all its parts are clearly written should you consider the manual ready to be reproduced in its final form.

Industrial manuals, of course, are more complicated than those you might design as a student. To show you how industrial manuals look, we have reproduced (1) the table of contents of a manual used in the aircraft industry and (2) the introduction of a manual in the field of source data automation.

(*1*)

CONTENTS

(2)

INTRODUCTION

General

The 2201 FLEXOWRITER automatic writing machine, PROGRAMATIC Model, is capable of many functions. It can sense codes in punched tape, edge-punched cards, or Hollerith coded tab cards to cause automatic typing of a document and selective punching of an additional coded tape or card at the same time. Furthermore, it can be manually operated to cause typing and code punching either alone or as a programmed part of automatic operation. Specialized auxiliary units may be cable-connected to provide additional tape and card input versatility to the writing machine.

There is a choice between two keyboards available for the printing of characters (manual or automatic). The single-case keyboard provides for typing of capital letters only, numbers, and special characters. The double-case keyboard provides, in addition, the printing of small letters and a wider selection of special characters.

A selection of readers and punches are available as component parts of the automatic writing machine. Two readers (code-sensing devices) operate with punched tape only or tape and edge-punched cards. A specialized tab-card reading unit is also available. Two punches (code-punching devices) operate to punch tape only or tape and edge cards.

Components

Five basic components make up the 2201 (see Figure 1): the reader, code translator, writing machine, code selector, and punch. Interaction of these components is as follows:

Reader. Mechanically senses codes punched in tape (edge-punched cards or tab cards) and converts each code into a series of electrical impulses which are sent to the code translator.

Code Translator. Converts the electrical impulses from the reader into a mechanical action of the writing machine to cause a keylever to be operated.

Writing Machine. Contains the power supply, the keylevers, and all necessary equipment to allow the 2201 to write a document and perform other machine functions.

Code Selector. When a keylever on the machine is activated, either manually or by action of the code translator, the code selector converts this action into a series of electrical impulses which are sent to the punch.

Punch. If the punch is on at the time these impulses are received, the code assigned to that keylever may be punched (depending upon other machine controls). This is the most common sequence of operation for the 2201. However, some functions may alter or bypass parts of this sequence. And all functions and operations of the 2201 are subject to programmed controls.

Operating Speeds

Reading (code-sensing) speed of the 2201 is 680 codes per minute. This speed will provide automatic typing at a rate in excess of 135 (5-letter) words per minute.

The keyboard can be operated manually to cause code punching at a speed of 1000 codes punched per minute, or approximately 200 words per minute. In addition, it will handle any two successive key-lever operations occurring at the rate of 1200 per minute.

Control

Specific operation of the 2201 is controlled from three areas: the keyboard, the removable control panel, and the field-switch rack (see Figure 2).

Keyboard. All manual entries and operator control of machine functions are handled through keyboard operation.

Control Panel. The 2201 is programmed for specific applications through codes sensed in the reader. However, the removable control panel is used to specify and modify machine functions and operation. This provides additional flexibility and versatility for all applications. Additional control panels are available to simplify application changes.

Field-switch Rack. This rack is removable and provides for the initiation or modification of machine functions at specific positions of the carriage. Additional field-switch racks are available.

EXERCISES

1. The following material was sent to the employees of a company. They complained, however, that it was too difficult to read. Turn the material into a set of formal instructions.

> It is felt that a reminder on our production parking lot regulation is in order at this time. The following plan was put into effect sometime past to make possible the quick and orderly exit of personnel at the close of business. Unless we follow this plan, much time is lost and possible accidents might occur.
>
> Between the hours of 7:30 A.M. and 8:05 A.M., the road leading to the parking lot will be a double-lane inbound route only and between the hours of 4:25 P.M. and 4:45 P.M. a double-lane outbound route only.
>
> Those cars which turn right in the direction of Butler at the main road will park in the first three rows beginning at the brookside of the lot.
>
> The balance of the cars which turn left at the main road in the direction of Boonton and Denville will park in the balance of the lot.
>
> On leaving at the close of the work day, the Butler bound cars will proceed in a line on the right side of the exit road. All others will proceed in a line on the left side of the exit road.

No person driving in the left lane will be permitted to turn right at the main road nor those in the right lane be permitted to turn left.

No cars will be permitted to enter or exit during the above periods contrary to the designated flow of traffic. Those persons who are dropped off or picked up by others who will not be parking their cars will do so at the main entrance located at the Administration Building.

2. The following directive was issued to certain employees of a railroad company. Many of them complained that the directive was difficult to read and to carry out. Rewrite in such a way as to answer the criticism.

CENTRAL DISTRICT YARDS

Effective January 1, 1970, a radio communication system will be placed in service in the Central District yards and in a base station located at North Belmont yard office.

The cabs of yard engines and engines assigned to transfer crews will be equipped with two-way radio service for use of employees in communicating with the Yardmaster at the base station who will have charge of yards in the district, including the station at Eggers Lane.

In addition to the above mentioned installation of radio equipment, an automobile Car No. KT-16 will also be equipped with two-way radio service for use by an Assistant Yardmaster operating from the base station at North Belmont to points throughout the Central District.

There has been placed at various yard offices in the Central District a small booklet entitled "Railroad Radio General and Operating Rules" issued by the Association of American Railroads, which has been adopted by the Federal Communication Commission, on August 25, 1965, and attached to its Order No. 126 pertaining to operation of railroads and radio equipment by railroad employees in connection with railroad operations. Employees using the radio service will provide themselves with a copy of this booklet and become acquainted with the information contained therein.

In making calls in accordance with Item No. 8 of the Operating Rules shown in this booklet, employees will sufficiently identify the station which they are calling and the station they desire to communicate with. For example—

North Belmont Yard calling Green Island Yard Diesel 3307
Central Yard Diesel 3008 calling North Belmont yard office
North Belmont yard office calling Car KT-16

Employees will notify the Yardmaster at North Belmont promptly when radios do not operate satisfactorily.

Please be governed accordingly.

3. Prepare a set of instructions to be used in one of these situations or in a situation of similar importance:

(a) In case of fire
(b) Sharpening a lawnmower
(c) Changing a tire
(d) Replacing parts in a radio or television receiver
(e) Operating a simple electric saw

4. Prepare a set of instructions to be used in operating a simple piece of apparatus found in a laboratory. The instructions are to be handed out to students performing a laboratory experiment.

5. Prepare a set of instructions for operating a simple machine or device used in the home.

6. In instructional form, write a description of a job that you held during the summer or which you are now holding.

7. Write a report to your instructor analyzing an instructional manual used with a machine or a process. It may be for operations, maintenance, or repair. Show the characteristics of the manual, especially in relation to the principles discussed in this chapter.

8. Select a manual used in connection with a simple operation. If you think that the manual is not clear, reorganize the material. By headings, show the various units, indicating what should be placed under each heading.

9. Gather data to be used in the preparation of a simple manual. Put the data in workable form and give it to another student to organize as in Exercise 5.

10. Write a short but complete instructional manual for a practical purpose. Some suggested uses are: for a student registering at your college; for building a simple device; for operating a machine in a laboratory.

11. The Bernoulli effect is described as follows:

> There is a very important relationship between the pressure at any point in a moving fluid and the velocity of the fluid at that point. The pressure in a rapidly moving fluid decreases as the speed increases. For example, if a tube is tapered, the speed of a fluid will necessarily be greater at the smaller end than at the larger; consequently, the pressure is less at the small portion of a constricted tube than at the larger portion.

Applications of the Bernoulli effect are seen in the venturi tube, in atomizers and sprayers, in the design characteristics of airplanes, in the curving of baseballs and golf balls, in sheets of paper held parallel with air blown between them, in the way a pipe of tobacco draws on a windy day, in the effect of a hurricane on the roof of a house, in the effect of a rapidly moving current upon ships anchored in parallel.

Write a direction sheet for a simple laboratory experiment illustrating the Bernoulli effect.

11. Proposals

Everyone entering a technical career will most likely write a proposal sooner or later. A proposal, as the term is customarily used in industry, is a sales piece designed to obtain work for a company. Its ultimate goal is to sell the services of a firm to some group that can use those services. In defense contracts, millions of dollars ride on a proposal. The proposal may be in the form of a brochure, a letter, or a long, complex report.

Through proposals you try to convince someone that you can do something for him, that you have the capacity to do it, and that you expect some support for the project and some recompense.

If you write a letter of application, you are really writing a proposal. You are saying, in effect, that you have certain training, skills, and potentialities, and that an employer or a company should take advantage of them and hire you.

When you apply for a fellowship or an educational grant of some sort, you are also writing a proposal. You are requesting financial support for training that may not benefit the donor directly, but will benefit the profession in the long run.

When an industrial concern or a research organization applies to a government agency such as the National Aeronautics and Space Administration for financial help in developing a program, it prepares a proposal. The National

Science Foundation, for example, recently received a proposal for a pollution study with the title *A Limnological Study of Lake George, New York.*

11.1 Kinds of Proposals

Some proposals are *unsolicited.* In other words, no one has told you that a need exists, but you think that it does and that you can propose something that will satisfy the need. For example, if you are a student, you may feel that the laboratory in which you work could benefit by having a procedures manual. Through a proposal you suggest that you and perhaps some others in your class could produce such a manual.

Other proposals are *solicited:* an individual or a group has decided that it needs a job done and it looks around for someone to do it. Perhaps the laboratory supervisor has already seen the need for a laboratory manual. He may ask his students to prepare such a manual. He calls on his experience and tells the students what kind of manual he wants. He supplies *specifications* about the length, the scope of the treatment, and the date when he wants the manual delivered to him, and he tries to get some money to pay for it. These are examples of proposals as they occur in fairly simple situations.

But proposals have become among the most important documents in the business world, in industry, and in research of all kinds.

Another example: the government decides that it needs a new electronic system for one of its space vehicles. Ordinarily the government is not in the production business. As a result, it puts out specifications on what kind of system it needs and how much it is willing to pay, and requests bids from those companies equipped to handle the matter. Perhaps the General Electric Company receives copies of the specifications, feels that it can supply the need, and thus starts out on a *solicited* proposal.

Such a proposal may run to hundreds of pages, involving dozens of engineers and administrators, and may cost thousands of dollars to submit.

Another type of proposal may involve a research group such as Battelle Memorial Research Institute in Columbus, Ohio. This research institute may have discovered it is on the verge of a breakthrough in the field of metal alloys, but that it cannot afford to use its own funds for further research. Battelle may then decide to submit an *unsolicited* proposal to NASA, describing its situation and requesting support for the experimental work.

Universities nowadays find themselves writing proposals not only for research grants but also for educational programs. When a college receives a grant from a foundation such as the Ford Foundation or the Carnegie Foundation, you can be sure a proposal of some kind has been written in advance.

You can see, then, that the proposal can be a simple document written on the undergraduate level or a highly complex piece of work important to the functioning of a company or group. *Unsolicited* proposals are likely to be aligned with research; *solicited* proposals with development and production.

11.2 Relation to Other Forms

The proposal borrows from other forms of technical writing. From the *formal report,* it usually borrows:

(1) An early statement of the problem or purpose.
(2) An early statement of important results or conclusions.
(3) A liberal use of heads and subheads.

From *technical papers* and *articles,* the proposal borrows devices that will attract attention and hold interest.

From *instructional writing,* certain proposals borrow the typical step-by-step arrangement and "how-to-do-it" technique.

From *technical description,* the proposal borrows the techniques of describing machines, processes, and theories, particularly for that portion often called "the technical section."

11.3 Proposals and Sales Letters

In Sec. 6.3.2 we stated that one kind of business letter is persuasive and is thus a sales letter. We use a job-application letter (Sec. 21.2.13) as an example. The technique of the sales letter can be applied as well to the proposal in letter form.

11.3.1 Attracting Attention

The first step in any letter, and therefore in any proposal letter, is to get the reader to read it, as we have pointed out in Chapter 6.

One way to accomplish this objective is to make the letter easy to read—by having a pleasingly simple format, good typing, short paragraphs, wide margins, and headings. Another way is by bringing yourself in contact with the reader through a common experience, such as a meeting both of you have attended, or an event in the news, or perhaps even a complaint from a third person—in other words, creating in the reader's mind a situation or a problem that he will recognize.

11.3.2 Creating Interest

The second part of the proposal letter is constructed around the situation or problem. Here is something that you are drawing to the reader's attention —the need to regularize a laboratory procedure; or a method for distributing service reports; or something more tangible such as the need for an air conditioner for the office.

11.3.3 Convincing the Reader

Once you have called the reader's attention to the problem or situation, then you must convince him that you can remedy the situation. For example, you may be able to demonstrate that you have devised a new laboratory technique, or a set of specifications for making up service reports; or that you have studied air conditioners and are able to compare the advantages and disadvantages of various kinds.

Convincing the reader also means that you can show him that you can benefit him by doing something efficiently and at a reasonable cost.

11.3.4 Obtaining Action

Your proposal letter should end by trying to get the reader to respond to your proposal, perhaps by appealing to certain standards such as ease of performance, efficiency, and cost, and by making it easy for him to get in touch with you.

To facilitate the four functions already discussed, you might consider these devices:

(1) Arrange material in a sequence conforming to the reader's idea of what is important. This technique, of course, will take some research on your part, perhaps a questionnaire or an interview.

(2) Anticipate objections early. Most people react more favorably when they realize that the proposer is willing to concede minor disadvantages, if any. They should be brought out in the open, refuted if possible, or admitted if necessary.

(3) Be more personal than is customary in a letter report, and do not be afraid to develop a relationship with the reader.

(4) Break material into clearly defined units by means of headings and similar devices.

An example of a proposal in letter form appears in Sec. 21.2.21.

11.4 Formal Proposals

Many proposals are not written in letter form. They are too complicated to be contained in letters, they are addressed to large groups, and they break down into a number of formal units. Proposals of this kind bear a considerable resemblance to formal reports.

11.4.1 The Approach

As it is easier to write a *solicited* proposal, one that has built-in specifications, we will concentrate largely on the *unsolicited* proposal.

Carl B. Palmer, director of research, American University, has this to say about unsolicited proposals:

> The unsolicited proposal, initiated by the researcher, is better suited for work near the research side of the spectrum than for development and production activities to meet defined requirements of the sponsor. Ideas cannot be programmed or specified in advance, but when a good one sprouts it should be cultivated and fertilized to ensure the maxi-

mum harvest. On the other hand, a new device will not be purchased, regardless of its merit, unless a need exists.

Accordingly, unsolicited proposals should apply to basic or applied research and lie in the areas of your greatest existing competence.

Clearly, the more time you can spend on analyzing the proposal material, the better the proposal will be. Familiarize yourself with all aspects that the proposal will cover. It is almost impossible to decide that a project is worthy of support without knowing the project inside and out. Discover who are working on it, how far they have progressed, and what they expect as a conclusion to their efforts. If it is the kind of project that has been designed to solve a problem, make sure that you understand what the problem is in the first place.

In other words, try to become an authority on all aspects of the proposal material. For more complex proposals, this advice is important, as you may have to present the proposal both orally and in writing before you obtain permission to carry it further.

These points described in the approach are ordinarily presented in abstract or summary form. Refer to Sec. 7.3. Recall that the abstract is a more or less linear reduction of a piece of writing and that the summary is a nonlinear reduction. Refer back to the quotation from Carl Palmer describing the principal kinds of proposals—one for research, the other for development and production. Although the distinction will not hold invariably, it does give you a hint about the use of the abstract or the summary in writing a proposal.

Like every other kind of reader, the recipient of a proposal will glance at the summary or abstract to see if it holds promise for him. It must be easy to read. It is well, then, to follow these principles:

(1) Put each point within a section in a format similar to the other points.

(2) Put the points in uncomplicated sentences and simple words. Let the sentences carry the message clearly and succinctly.

(3) Make the sentences in a paragraph and the paragraphs in the summary as parallel as possible. Parallelism will result in clear statements.

11.4.2 Presenting the Solution

Describing the solution that you have in mind is usually the most difficult part of the proposal to write, often because you have been unable to test your ideas for lack of the very resources that you are requesting. However, the solution is the meat of the proposal. Without it, you have nothing to sell to the proposed sponsor.

It is a good idea to summarize the solution just as you did the approach. In Chapter 4, we have referred to a basic technique in exposition—that is, to describe from the general to the particular and from the abstract to the particular. Do this as your summary of the solution.

The summarized solution should be followed by a more detailed description. Keep your reader very clearly in mind, make your points logically, and be as realistic as you can in your evaluation of the proposal situation. The reader should be able to think of you as a person who has his facts straight and knows what he is doing.

11.4.3 State-of-the-Art

At this point, some proposal writers suggest that you should write a section called *state-of-the-art*. If you will consult Sec. 6.2.5, you will see that in progress reports, you should describe what has been done on a project in the past.

The *state-of-the-art* section has a similar function. It serves several purposes: (1) it brings the reader up-to-date on the material of the proposal; (2) it points out the contributions that other people have made; and (3) it shows that the writer has already done some research and documentation on the subject. This section should increase the reader's confidence in the writer.

The *state-of-the-art* section is particularly important in unsolicited proposals. In solicited proposals, however, the customer or client will have initiated the proposal and should know the background of the material.

11.4.4 Technical Section

The technical section is where the proposed research is described. In proposals issuing from large companies, this portion coordinates the work of several engineers and scientists who will be doing the research and development. This part of the proposal is often given to a professional technical writer to meld into a unified statement.

The technical section may require a description of machines, processes, and theories. If so, consult Sec. 4.5, *Suggested Outlines,* not to follow them slavishly, but to adapt them to your particular needs.

In addition, we suggest that you also consult the chapter on instructions, particularly Sec. 10.4. Just like the manual, the technical section may deal with both informative and instructional writing. In this context, be sure that you use headings advantageously.

Although we cannot tell you what you should cover in the technical section of any one proposal, one authority says that a broad outline might look like this:

(1) Proposed program and schedule.
(2) Fundamental design considerations.
(3) Deviations from specifications—if the proposal is solicited.
(4) Some proof of the reliability of suggested procedures.

11.4.5 Resources

Like the sales letter, the proposal must back up its claims—by citing capability, personnel, and facilities.

In terms of an individual, *capability* means that you have the stamina to complete the task implied in the proposal. It also means that you have the ability and talent to do so. You may find it necessary to prove that you have these resources through an interview with the sponsor, by citing any past experience you have had, or by obtaining references from other people.

In terms of a company or organization, this section should deal with past experiences in the area of the proposal in order to demonstrate that it has the business acumen to organize and carry out the project and that it has a well-earned reputation in the industrial community.

A small company with little experience and unknown personnel will, of course, meet strenuous competition from companies with wider experience and longer work records. In the long run, the projects proposed are only as impressive as the *personnel* who will be working on them. In a simple proposal, it may be enough to know that the proposer himself will be carrying out the project and that he is capable and experienced.

Simple statements of this kind, however, will not be sufficient for complex proposals. If an agency or a company is going to spend millions on a proposed research idea or on a potential product, it must know who will be the overall director, the section supervisor, and the key personnel.

A section, then, should be devoted to a description of each of these persons in these terms:

—education
—professional accomplishments
—professional honors
—research carried on in the past
—proposal projects with which an individual has already been associated
—papers, articles, and books he has published and important professional speeches he has given

The description of *facilities* will take two directions. First, the reader will be interested in knowing what are the existing facilities, ranging from laboratory equipment on hand to physical plant layout. Second, the reader will be interested in knowing what additional facilities will be required in order to carry out the proposed work.

Facilities, whatever their nature, must relate directly to the project and must be described accurately. If they are to be acquired, costs should be enumerated and later tied in with the cost analysis.

11.4.6 Cost Analysis

It would be difficult in this space to give all the fine points of cost analysis. In solicited proposals, various items are usually clearly specified. In unsolicited proposals, the proposer must take particular care that all items of expense have been anticipated.

In general, costs will give a very good indication of what you have in mind. A run-through of some typical proposals discloses the following as important elements in this section:

—direct materials
—equipment already purchased
—special equipment
—engineering labor and overhead
—manufacturing labor and overhead
—travel
—publications
—professional consultants

11.4.7 Suggested Outlines

From various industrial sources have come the following three outlines. In spite of some differences, they show that there are certain standard items that must go into a formal proposal.

I. Approach
   Description of the Solution
   Discussion
   Resources
   Costs
   Time of Completion

II. Summary
   Introduction and State-of-the-Art
   Proposed Research
   Costs
   Personnel

III. Summary
   Introduction
   Statement of the Task
   Technical Section
   Management Section
   Capabilities and Experience
   Personnel
   Facilities
   Costs

An example of a formal proposal is reproduced in Sec. 21.2.22.

EXERCISES

1. You and a fellow student feel that you are capable of starting a tutoring service for incoming freshmen who may find it difficult to adjust to a college curriculum. Write a letter to the dean of students or some other appropriate administrator, asking for sponsorship of the proposal.

2. As the representative of the student body of your college, prepare a proposal for a potential sponsor for support of a study to determine the feasibility of a small cultural center for your campus.

3. If possible, obtain from the research division of your college a proposal that has been prepared for a government agency. Analyze and evaluate the proposal and write a short survey report on your findings.

4. Prepare an outline for a proposal to a foundation for a grant to support field trips in connection with a course in your curriculum. The trips would be under the supervision of faculty members. Without completely writing the proposal, indicate major section headings and provide summaries of the content of each section.

5. The publications department of a large company is in need of additional equipment. Three requests have been made to the purchasing office with no results.

Now the following letter has been sent. In its present form it can be criticized for format, organization, and clarity. Turn this letter into what you consider to be a more acceptable proposal.

Mr. H. M. Farnsworth
Purchasing Director
Miles Aviation Company

Dear Mr. Farnsworth:

In April of this year three purchase requests were submitted involving an expenditure of approximately $7,300 for duplicating and binding equipment for use in the Research and Development Center. In the interim period, many studies and surveys have been made relative to the advantages, as well as the logic in buying this equipment.

To date, no decision has been made relative to these purchase requests (number A-206077, A-206078, and A-206079). With the volume of research and development publications increasing rapidly, the cost savings possible with this equipment becomes of greater and greater significance. If it had been possible to install this equipment on the first of May 1969, conservative estimates indicate that the savings during the past six months would have been in the order of $12,000 over outside printing costs.

In spite of the dramatic cost savings possible, it is emphasized that the prime need for this equipment is to facilitate the duplication of a large variety of technical reports, proposals, and other data which are generated in support of R&D activities. It is not economically feasible to have all of these printed outside, yet neither is it deemed wise to sacrifice quality of duplication when these data are prepared for our customers or prospective customers. To give you an idea of the volume of work falling into the above category, the R&D Publications Group has turned out so far this year 94 reports of various types, 9 brochures, and 4 technical society papers.

It is very logical to ask why this equipment is needed in the Research Center and under the control of the Publications Group. The answer to this lies in the fact that R&D publications are produced on a high-volume, short-deadline basis. To understand the production concept, it is necessary to point out that the three IBM typewriters now operated by this group enable final typing to be performed on "Colitho" paper plates. These plates can be placed directly into an offset duplicator, such as that requisitioned, and 50 copies obtained within two minutes. In preparing a 25-page report, the pages could be typed and run off immediately so that a completed report could logically be obtained within *an hour* after the last page is written. This allows time for review by the responsible technical groups and for corrections and revisions as may be required. In sending work to an off-site printer, the flow time is in terms of days and weeks, thus frequently eliminating review time and forcing distribution of the report without proper coordination.

The items requisitioned are all small, "table-top" sized units with the exception of the duplicator and collator. These latter machines are strictly "office-type" equipment and illustrations are provided herewith to give you an idea of their size.

The Publications Group is already up to full budgeted strength and is working to full capacity. In order to handle the additional volume of work once the laboratories become operational, some time-saving methods must be introduced. The purchase of this equipment is the logical step and will greatly streamline the entire publications effort. No additional manpower will be required. The two men operating the ozalid machine can run this equipment.

I wish to personally urge your favorable considerations of these purchase requests. Dr. O. L. Drysdale, Division Chief, has already indicated his endorsement by personally signing the purchase requests which were submitted.

Sincerely,

J. M. Dekker, Manager
Publications Department

6. The training division of a telephone company would like to have its resources used to a greater extent by the engineering division. Currently, the training division is promoting two courses:

A. Introduction to Plant Engineering

—Intended for new employees.
—Would be taught by the local district plant engineer at local branches.
—Covers organization of local district offices, location and organization of files and records, practices, and agreements.

—Discusses area and division plant engineer's organization and plant engineer's relationship to other groups and departments.
—Cost: $700 for a one-week, 20-hour session.
—Limited to 20 people.

B. *Basic Plant Engineering*
—Intended as a beginning course for new people and as a refresher course.
—Takes 4 weeks of 10 hours each.
—General items of instruction include cost studies, accounting, right of way, joint use, transmission, protection, records, and field note preparation.
—More specific items include extensions of pole line and wire, subsidiary conduits, and buried wires and cables.
—This is a survey course; more exhaustive treatment would follow in an advanced course.
—Instructors: specialists in each of the main subjects.
—Location: company headquarters.
—Cost: $1500.
—Limited to 40 people.

Write a proposal to the Director of Engineering Services, urging that his division take advantage of the courses offered by the training division. You may add to the information already provided.

12. Technical Papers and Articles

The engineer or scientist who has discovered or developed something new is likely to report his work in a paper or an article in one of the technical magazines. In general, writing of this kind is governed by the same principles as other forms of technical communication, and almost everything said in this book is applicable.

The term *technical paper* usually describes a piece of writing that is ultimately destined for a journal published by a professional society. Often, it starts out as a manuscript to be presented at a meeting, after which it is printed with little or no revision (but see Sec. 13.2.1 on the reading of speeches).

The term *technical article* is more general. Writing of this kind is usually done directly for a technical magazine put out by a publishing company. Sometimes the readers are not so technically trained as the writer; the article then must become semitechnical in nature. Sometimes the information can be carried to an even wider group of readers in a popularized article.

12.1 Kinds of Periodicals

In this country there are many technical publications. We have seen the number given from 5,000 to 10,000; it all depends on the definition of the term *technical*. A practicing engineer or scientist can increase his prestige by writing for some of these publications. But the engineer more often writes with another purpose—to keep the reading public aware of what both he and his company are doing. The technical article then becomes very important as part of the public-relations program.

A student, too, may be able to write for technical magazines, for many college writing courses have been designed with the very realistic purpose of teaching the student how to write for a real audience. Don't overlook the journals and technical magazines. You will find student articles featured in magazines such as *The IEEE Student Journal* and *The Bridge of Eta Kappa Nu,* and from time to time in magazines such as *Chemical Engineering, Research/Development,* and the *American Engineer.*

In addition, you should consider the possibility of writing for your own college publications. There are at least sixty student engineering magazines in as many colleges. They offer an excellent opportunity to practice technical-article writing.

Technical publications are divided into three general groups:

(1) *Journals sponsored by the professional societies.* Representative of these are the *SAE Journal* published by the Society of Automotive Engineers; *The Journal of Chemical Education* by the American Chemical Society; and *Civil Engineering* by the American Society of Civil Engineers. Journals of this kind contain papers written by specialists for specialists in the same field.

(2) *Technical magazines.* Typical ones are *Product Engineering, Machinery, Electronic Design, Machine Design,* and *Factory.* Because these magazines are commercial and are published by profit-making companies, they are inclined to be more interested in articles dealing with applications of new knowledge rather than with theories, and each magazine attempts to appeal to a distinct group of readers. Because practical matters are stressed by the technical press, these publications require a somewhat different kind of treatment than do the journals. Devices, processes, techniques, and applications make up the bulk of the subjects discussed, together with surveys of the broader aspects of technology.

(3) *House organs and company magazines.* These differ from the other two groups only in that they originate in industrial companies. *Electrical Communication,* put out by International Telephone and Telegraph, strongly resembles a professional journal. *Westinghouse Engineer* is broader in scope and resembles a technical magazine. Most papers and articles for company magazines come from company employees.

12.2 Preparation for Writing

12.2.1 The Reader

Like other communications, the technical paper and the technical article must be prepared with a definite group of readers in mind. In writing the technical paper, for example, ask yourself these questions: Will your readers be specialists in your own field, with a full knowledge of its problems, its conventions, its language? Or will they be predominantly technically trained people with other specialties, who are conversant with the fundamentals of science but need instruction in some of the elements of your particular field? In general, the readers of technical *papers* are more interested in basic principles and in new ideas and experiments than they are in practical applications.

The readers of technical *articles* usually want to know what is going on in the technological world. They are first of all interested in facts, rather than in theories. They will also want to know how to apply such information. They will expect the writer to stress the usefulness of whatever is described, be it a device, machine, or process, and especially to show its practical application to their needs.

12.2.2 The Material

The material for technical journals is as broad as human knowledge and at the same time as specialized as the scientists who write it. Laboratory experiments, theses of all kinds, learned speeches delivered at professional meetings —from these come the many and diversified topics adapted by the journals.

Technical magazines, on the other hand, will be more restrictive in the material they accept. Each magazine has its own needs, its own requirements, and its own concept of what its readers want. Consequently, ask yourself the question: what kinds of topics seem to create the greatest interest in editor and reader?

Among subject classifications appearing in the technical magazines, these head the list:

Machines and devices
Processes and procedures
Plant descriptions
Operation and performance
Maintenance
Safety
Increased efficiency
Costs
Personnel problems
Technological trends

12.3 Writing Papers and Articles

12.3.1 The Technical Paper

The construction of a technical paper is essentially the same as the construction of a report, but the paper usually consists simply of one main presentation, in logical order, plus a short abstract. (See Sec. 7.3.2 for suggestions about the abstract.) Although the technical paper does not contain all the separate elements of the formal report, it can practically always benefit by being broken up into logical sections with headings and subheadings (usually of not more than two degrees).

Many technical papers seem to follow this outline:

(1) Abstract.

(2) Purpose. Explains, as in the report, what the author is doing, and why, and in general orients the reader to the topic under discussion.

(3) Conclusions. Again, as in the report, the reader is entitled to get his principal information at once.

(4) Procedure. Tells how the project was carried on, with all the detailed steps coming after the broad, overall view.

Another outline that is finding increasing favor is:

(1) Abstract
(2) Introduction
(3) Results
(4) Discussion
(5) Experimental (set in smaller type when printed)
(6) Conclusions or Summary

In writing for professional journals, contributors should refer to any style manual provided by the journal, such as those listed in Sec. 21.3. Most libraries should have them. Here is the required outline for articles found in the style manual of The American Society of Mechanical Engineers:

(1) Title
(2) Author's name, business connection, and mailing address
(3) Abstract
(4) Body of paper
(5) Appendixes
(6) Acknowledgments
(7) Bibliography
(8) Tables
(9) Captions for all illustrations
(10) Photographs and other illustrations

12.3.2 The Technical Article

The technical article, slanted for a more commercial and business readership, is likely to follow this outline:

(1) Lead
(2) Lead-in
(3) Body
(4) Ending

We will consider each one of these parts in turn.

(1) Lead

The lead is usually one paragraph in length; it is the first thing read; and it has the dual function of introducing the reader to the subject in a general way and of attracting attention. The kinds of leads given here are samples of those most frequently used.

The offer of something new:

Aluminum is used instead of copper for electrical conductors because it offers opportunity for reducing cost and weight. . . .

Summary:

Broadly speaking, dimension control has two main uses in design and in manufacture. . . .

Historical:

Graphical symbols are one of the oldest forms of written language known to man. The ancient caveman drew crude pictures of things about him. . . .

Comparison of the old and the new:

Although the potentialities of computers were recognized early in the nineteenth century, large-scale digital computers are a recent development. . . .

Reference to authority:

Vance Packard in his book *The Hidden Persuaders* tells of an experiment aimed at finding out the effect of color in merchandising.

(2) *Lead-in*

The lead-in is the stepping stone by which the reader gets to the main portion of the article. Some authorities call the lead the point of *general interest* that attracts attention and the lead-in the point of *specific interest* that holds attention.

The lead-in tells you specifically what the rest of the article is about. Thus it is a statement of the *unifying idea* of the article.

We consider the unifying idea one of the most critical phases of writing an article. In short, it is the core of the article. It is what your article is about. Everything in the article must contribute to it, and anything that does not should be ruthlessly cut out. Think less in terms of the formal outline and more in terms of the unifying idea. Say to yourself: what am I writing about and what do I want the reader to get out of it? That is the *unifying idea.* Present it as part of the lead-in.

In most technical articles, especially those read almost solely for their informational value, the lead and the lead-in overlap and may be contained in a single statement.

Imagine that a lead sentence or paragraph precedes each of the following lead-in statements.

Statement of specific use or idea:

One of the neatest and fastest-growing approaches to the solution of some of these problems [efficient heat transfer and corrosion prevention] is the use of volatile, film-forming amines. . . .

Summary:

We can best determine the true importance of the engineering staff by reviewing its objectives, its relationship to the rest of the company, and its contributions. . . .

Explanation of purpose:

This article presents, as comments from both members of a typical team of engineer and technical writer, the most frequently voiced of these principles. . . .

Here are some examples of the lead and lead-in combination.

Lead (reference to authority)

A significant provision of the patent statute is: "Subject to the provisions of this title, patents shall have the attributes of personal property. Applications for patents or any interest therein shall be assignable in law by an instrument in writing."

Lead-in (statement of specific idea)

On this provision rests not only the right of an inventor to sell or transfer his patent, but the collateral right to license its use to others while retaining ownership of the patent and its attendant monopoly.

Lead (summary)

To promote participation in technical society work, the governing board should create an encouraging climate and an atmosphere of enthusiasm, promote individual activity among its members, assist engineers, and publish information of specific interest.

Lead-in (purpose for writing)

We have repeated these basic principles for one purpose—to call attention to the apathy that exists among the memberships of some societies and to give some ideas for combatting it.

Lead (comparison of new and old)

The old "sourdough" prospector scrambling over outcroppings with his hammer, compass, and magnifying glass is rapidly becoming legendary. A few may still exist, but they are being replaced by men of science who peer into the untold resources of nature with the lenses of their own inventiveness.

Lead-in (statement of specific idea)

Perhaps the greatest advance in recent years has been in the use of aerial surveys, where, with the latest instruments, up to 80 percent of a prospective area can be eliminated in a very short time.

Lead (something new)

A totally new concept in systems drives has been developed and applied successfully in a number of industries. It consists partly of a basic reorientation of application and design engineering and partly of a standard set of modular power and control components.

Lead-in (statement of specific idea)

Called the T-100 concept, it benefits users directly through improved system accuracy and response, faster shipments, lower costs, versatility, quick efficient startups, and easy maintenance.

(3) *Body*

The body of the technical article must be based on solid organization. All the devices and techniques already discussed apply to this main portion. In addition to deciding what the unifying idea will be, be sure that you plan ahead and make an outline. Magazine editors can see through a poorly organized piece of writing.

If you take the time to look over material already published, you will find that three methods of organizing the body are at the top of the list.

(a) *Chronological.* As discussed in Sec. 2.7, the chronological method presents material as it occurred in time. First things are discussed first. This method is particularly adaptable to the description of processes. However, we must inject a word of caution. The chronological outline is frequently more appealing to the writer than to the reader. Visualize the order of the information from the reader's point of view. Sometimes a chronological order can make deadly reading.

(b) *Importance.* This second plan is based on the assumption that certain parts of a machine or steps in a process are more important than others. Because the interest of the reader may wane as he reads the article, it may be wise to discuss the most important elements toward the beginning of the body.

(c) *Climactic.* Some subjects, especially those of a persuasive or argumentative nature, lend themselves to a treatment that proceeds from climax to climax. In this procedure, you may make a point, then go on to establish others through logical arrangement. The article thus contains a series of high points that establish certain principles.

To outline the body of your article is not enough. You must always imagine the reader at the other end and learn how to hold his attention at the same time as you give information. These devices are commonly used:

Statistics. Statistics give authority and proof to our words. But don't use statistics entirely by themselves, in the form of figures and tables only. Put them in situations involving people and places.

References from the experts. Interpret your information in terms of people of ability and reputation. We must recognize, however, that reference of this kind is not necessarily proof. It is simply another way of lending weight to your ideas.

Incidents. Readers can better visualize an abstract piece of information when it is put in terms of description and narrative. An incident is something happening; describe it in terms of what happened in the laboratory, who worked on the project, unusual occurrences, and the place and time involved.

Comparisons and analogies. When the material is unfamiliar to the reader, translate it into terms of things he already knows.

(*4*) *Ending*

Many writers find it difficult to end their articles (see Sec. 2.7.3). Think of the ending as making a full circle with the beginning. For example, if it was necessary to begin the article with a summary, it may be as necessary to end with a summary. These are some of the endings most commonly used:

Summary
Logical conclusion
Application of basic ideas
Forecast of future events

12.3.3 The Semitechnical Article

You may find yourself writing for a more general audience than those who read professional journals and technical magazines. Magazines such as *Popular Science* and *Science Digest* come to mind. And some of the company magazines, because they advertise their companies as well as give information, tend to favor more popularized articles.

The basic techniques for semitechnical articles are those used in technical articles. But the following devices should help to make your material clearer to the untrained reader.

(1) Consider using other leads than those already cited. The more popularized the article, the more appropriate these leads:

(a) Narrative beginning:

> When Cheops of Egypt built his tomb, the Great Pyramid, he used gypsum plaster to hold the giant stones together. From then until the time of the ancient Romans, gypsum was used primarily in construc-

tion and ornamental plasters. Then, one day a fire broke out in Rome in a building that contained gypsum plaster. The fire burned for a while, but went out when it reached the walls of the building. Naturally people were puzzled, but some experimentation showed that the gypsum plaster had the ability to limit the spread of fire. Another use for gypsum had been found.

(b) Description of people and places:

Charles Montague Doughty was the most obstinate of men. His bullheadedness led him to disregard the wise admonitions of men who knew better to stay away from the then little-known Arabia.

(c) Dialogue and human interest:

At Oregon State College, we hadn't seen the sun in two months and the persistent drizzle was changing to snow.

A group of us students were lacing up our boots in the basement of the forestry building when a visiting British professor and his guide came down the stairs. He noticed us and asked the guide:

"Are you having some work done around the buildings?"

"No. These are just students getting ready to go outside for a class."

"Surely not in this weather." The visitor was amazed.

"These are forest engineering students. When they leave college and get out on the job, they will have to work outside under all conditions. This is part of their training."

(d) Startling statement:

If world population continues to increase at its present rate, according to a prominent demographer, within *x* years the whole surface of the earth will be solidly covered with human beings. Within *y* years, this human sphere will be expanding with the velocity of light. And *x* and *y* are much smaller numbers than you would ever believe.

(2) Write very short sentences and paragraphs.

(3) Use short words with action and visual appeal; avoid technical terms.

(4) Provide headings.

(5) Write to the eye as well as to the ear. Use listings, tabulations, underlining (or italic type), capitalization.

(6) Quickly identify things by their shape, size, color, material, texture.

(7) Make statements in a positive manner.

(8) Use repetition to emphasize points.

(9) Personalize your writing. The farther you move from pure exposi-

tion, the more effective it is to put yourself and other people into your writing.

(10) Don't try to be unique in your methods of presentation. As in other forms of reporting, the semitechnical article gives information.

12.4 Illustrations

You will, of course, look over copies of the magazine for which you are writing, especially to see if it uses illustrations to go with the text. Most technical articles should be supplemented by illustrations and other graphic material. You will be able to obtain your illustrations from various sources: photographs you have taken, pictures from books and company sales literature, graphs and tables that you will yourself provide. Be sure that you ask permission for the use of illustrations from outside sources and that you give credit for them.

You must provide captions for all illustrations. Unless a caption is an integral part of an illustration, as in a graph or chart, do not attach it permanently to the illustration. You can ruin a photograph by typing or writing on it. Moreover, an editor may want to change your caption. Type it and attach it to the illustration with paper clips or rubber cement. See Chapter 19 for tips on illustrative material.

An article written by a student appears in Sec. 21.2.23. Your library has many examples of technical papers and articles. Look them up in the journals and magazines and inspect them critically. Here is a list for you to examine for examples of publishable articles:

Allen, Durward L. "Natural Resources and the Cult of Expansion," *Conservationist* (December 1967–January 1968).

Anklan, Deane R. "County Engineer-Consultant Relationship: How It Works Best," *Better Roads* (June 1969).

Carpenter, David. "Machines Made to Measure," *World Health* (March 1969).

Chadwick, Wallace L. "Contracting Conditions and Construction Quality," *Civil Engineering* (August 1965).

Christy, R. T., and Kistler, D. R. "The Operation of a Business Information System," *Sperry Rand Engineering Review* (Vol. 22, No. 1, 1969).

Colton, Willard. "Chemical Stepping Stones," *Steelways* (January–February 1969).

Davis, W. Jeff. "Camille Hits Gulf Coast," *Ocean Industry* (October 1969).

Elonka, Steve. "New York City's Tough New Air Pollution Bill," *Power* (August 1966).

Findlay, John W. "The Scientific Challenge of Space," *Astronautics and Aeronautics* (February 1968).

Golomb, Solomon W. "Extracting Signals from Noise," *IEEE Student Journal* (September 1964).
Laird, Anna K. "The Dynamics of Growth," *Research/Development* (July 1969).
Leary, Frank. "Who's Winning the Undersea War?" *Space/Aeronautics* (February 1969).
Moon, W. D., and Weiner, R. J. "Processing Seismic Data in Real-Time," *UnderSea Technology* (October 1969).
Moyes, E. R., et al. "New Problems Challenge Engineering Management Today," *SAE Journal* (March 1968).
Nims, A. A. "All-Weather Mapping by Radar," *Westinghouse Engineer* (May 1968).
Plant, Robert. "The Dangerous Atom," *World Health* (January 1969).
Schon, Donald A. "The New Regionalism," *Harvard Business Review* (January-February 1966).
Shoemaker, A. F. "Strengthening Glass and Glass-Ceramics," *Mechanical Engineering* (September 1969).
Smith, Bernard. "Ideal Urban Transit," *Astronautics and Aeronautics* (September 1968).
Svensson, E. L. "The Lunar Television Camera," *Westinghouse Engineer* (March 1968).
Walmer, Warren L. "Venting Pressure Flare-Ups with Rupture Discs," *Machine Design* (August 29, 1968).
Whitman, Kirwin. "Organizing the Program for Control," *Chemical Engineering* (December 5, 1966).

EXERCISES

1. Write a short report of the article requirements of a technical magazine, based on an examination of at least one article from each of three issues of the magazine. Some typical magazines are: *Chemical Engineering, Product Engineering,* and *Machine Design.*

2. A technical magazine gives these specifications: "We will accept articles of general scientific interest which cut across technical fields and are understood by engineers and scientists. We want articles 500–700 words that deal with things or ideas that the average technical reader is likely to come across.

"We do not want articles on topics of too specialized interest. Because we cannot use too many illustrations, the articles must be lively and appealing to our readership."

Choose a topic and describe how you would organize the material to fit the above specifications.

3. The selection given here is a brief article to be published in a technical magazine. Without changing the content or the organization of the selection, rewrite it to improve general style.

REACTOR CONTROL FOR CONSTANT CUTTING SPEED

In the case of each metal there is a best machining speed. To operate in excess of this ideal speed results in damage to the tool. Machining at a slower rate causes decrease in production. It is therefore obviously a problem to maintain constant cutting speed when machining the face of a jet-engine disk, which is made to rotate as the tool is moved across it in a radial direction. It is clear that the feed of the tool must inevitably change as the tool moves outward from the center or inward from the circumference. A generally used method of effecting approximately the desired speed control has been the employment of a mechanical linkage to a rheostat in the motor-field circuit. This method has, however, proved to be rather clumsy.

Satisfactory to a much greater degree is a solution involving the use of a control in which a small reactor is incorporated. A simple, small-sized cam is cut for each size of part that is to be machined, and also for each type of part. A probe from the reactor, which is contained in a box about four inches wide by four inches deep by four inches high, keeps riding on the cam, and moves up or down in a vertical direction —a maximum distance of not in excess of one half of one inch. It causes the motion of an armature that changes the reluctance of the magnetic circuit in the reactor. An electronic exciter makes the motor field strength to vary and results in the d-c motor speed being varied to make the cutting speed constant.

The motor is capable of operating within a speed range of from eight to one, the variation being obtained solely by control of motor field strength. The doubled speed range is because of two sets of windings per each of the four poles connected in different sequences, obtaining net flux changes much larger than normal.

The system obviates the need for mechanical linkages. Only one step is involved. Also, the device makes use of no sliding contacts or rotating bearings.

4. Select a paper from each of three successive issues of a technical journal. Analyze these papers in order to determine the requirements of the journal for form and presentation. In report form, record your survey and analysis.

5. Make a similar analysis of three articles from a technical magazine. In addition, come to some conclusions about the differences between writing for journals and writing for magazines.

6. Select a technical paper and write an informative abstract of it.

7. For a projected article for your college magazine, write a short report analyzing your own fund of knowledge, the interests of your likely readers, and the methods by which you will present your material.

8. For purposes of practice only, select a technical or scientific subject and write the beginning of an article as it might appear in a journal, a technical magazine, and a general magazine.

9. Select a technical magazine. Prepare a short article for publication describing a new product, a new procedure, or a significant trend in technology.

10. Prepare an outline for a technical paper or article. Justify your outline by referring to the material with which you will be dealing.

11. Prepare a short research paper to be submitted to a professional journal. Base the paper largely on library work and documentation.

12. Before you start writing the rough draft of a technical article, state the devices you will use to lend weight and authority to the body of the article. Submit your statements in numbered and listed form.

13. A description of the Bernoulli effect was given in Exercise 11 of Chapter 10. Use the same material to write an article of not more than 500 words for a general magazine on how the reader may observe the Bernoulli effect in action or how he may perform a simple home experiment without laboratory equipment.

13. Oral Reports and Speaking in Public

Every time you discuss your work with your boss or one of your colleagues, you are making an oral report. Most engineers and scientists do this sort of informal technical reporting very well. But when you write a paper for one of the journals, you will probably first have to deliver it orally at a meeting of a professional society. This more formal sort of oral reporting calls for greater skill and more planning than everyday informal conversations about technical matters. The better known you become in your profession, the more likely you are to be called on to give other kinds of talks and speeches. In addition to the oral report before a professional society, you may find yourself speaking at business conferences, taking part in panel discussions, and giving nontechnical speeches in your community. In this chapter, which is far from being a complete treatise on public speaking, you will find some pointers that should help you to do an effective job whenever you have to talk before a group of people.

13.1 Material for the Speech

13.1.1 Technical Speeches

If you have attended meetings of the professional societies, you have probably discovered that all too many of their speakers simply stand up and read the papers that they have written for publication in the journals. This is poor practice for a number of reasons. Most people can absorb knowledge more readily through their eyes than through their ears: when they read a paper they can assimilate far more complicated ideas than they can when they listen to a speech. Therefore an oral report must in general be *simpler* than its written counterpart.

Furthermore, the listener cannot turn back and reread the passages that he has missed or that he has forgotten, and he cannot pause to mull over particularly difficult ideas. Therefore the oral report must not only be simple; it must supply frequent reminders and restatements of key points. Consequently it must contain considerably less information than a written report of the same length.

Finally, written language and spoken language are quite different. As a result, the words that look natural and appropriate on paper sound unnatural and inappropriate to the ear. Spoken language is, naturally, easier to listen to than written language.

For all of these reasons, the written paper is very seldom suitable for oral presentation. Any time you have occasion to deliver a formal talk, by all means take the time and the trouble to prepare a version of your material that fits the needs of listeners, not readers.

Some societies ask their speakers to give 10- or 15-minute oral presentations of papers that might take 30 minutes or more to read through. If you condense a 30-minute paper to a 10-minute talk by linear reduction—that is, if you omit none of the ideas but only some of the explanations—you will present your hearers with a concentrated dose that they will be unable to understand even if they want to. Any time you have to prepare a short oral version of a longer paper, pick out the two or three most important points of the paper, and develop them thoroughly and leisurely. Refer your listeners to the printed paper for the minor points if you want to; but otherwise, exclude these matters rigorously from your talk.

13.1.2 Presentations

The word *presentation* has entered into the engineering vocabulary; it signifies a special type of technical speech distinguished by the liberal use of visual and audio aids. A presentation is likely to be given before a group of experts. Speakers who make presentations are usually called on because of their first-hand knowledge of a specific field. They are likely to be engineers first and speakers second.

Presentations are an integral part of business communication. Like most speeches, they must be well put together and they must be rehearsed and stage-managed to a certain degree. Most presentations rely heavily on a written text, or at least on complete notes, for presentations will be incorporated into reports, minutes, and other industrial publications.

If you are a student, you may give a presentation of your senior thesis. Later, as a practicing engineer or scientist, you may be called on to give a presentation in support of a proposal, or as your contribution to a seminar or a committee meeting.

The speaker should attempt to use the techniques discussed elsewhere in this chapter. He should also think in these terms:

(1) *Purpose of the Presentation.* What you will say and how you will say it depend on whether your sole purpose is to inform your audience, to instruct it, or to persuade it. Of course all three motivations may be found in one presentation.

(2) *Kind of Audience.* A technical audience will dictate one kind of approach and a semitechnical audience another. You may be speaking before a group of professors all in one discipline, or you may be scheduled to speak before a mixed professional group. A management group will be more interested in administration and supervision; a research group in theoretical considerations and in tests.

(3) *Facilities.* Unless you are a particularly poised and experienced speaker, you may be considerably shaken at the time of your presentation because you have not seen in advance the room in which you are to speak, the lectern (or lack of one), or the facilities for various media, and have not received some idea of how near the audience will be to you.

We have seen a well-prepared presentation almost ruined because every contingency had been taken care of except the lighting; the speaker couldn't read his notes and the audience couldn't see the speaker.

A presentation, then, should be carefully prearranged to provide lectern, lectern lighting, facilities for visual aids, stands for charts, and other support material.

(4) *Visual Aids.* The care taken to supplement the speech with visual aids characterizes the professionalism of the presentation.

In this age we have been conditioned not only to hear the word, but also to see it: to have material rendered in terms of pictures, illustrations, charts, and graphs. Therefore, the presentation should pay particular attention to these aids and the mechanical means of offering them.

Slides are easy to use, and most companies and colleges have the apparatus to show them. But they do have the built-in disadvantage of requiring a minimum of room lighting, thus making it difficult for an audience to take notes.

The opaque projector is most useful for placing spur-of-the-moment material before the audience. It will project handwritten as well as typed or printed material without elaborate preparation. But it, too, has disadvantages. It requires a darkened room, it does not supply much enlargement, and it generates a lot of heat that may be injurious to some materials.

The overhead projector has a variety of uses, but most of the material to be projected must be prepared in advance in the form of full-sized transparencies. Nevertheless, it has advantages over the other machines. It does not require a dark room, the speaker can write directly on a special plastic roll while he is speaking, and the projector comes with a number of useful attachments.

Flip charts are charts joined at the top, which, as the name implies, can be flipped back and forth over a tripod. If the lettering is large and clear, the material not cluttered, and suitable colors used, flip charts are very useful in presenting charts, graphs, and tabular lists.

Regardless of how the visual aids are used, you should test them against these principles:

They should be keyed in with the oral part of the presentation.

They should be easy to use. It is possible for a speaker to make it difficult for himself and his audience by cluttering up his speech with too many aids.

They should illustrate the main points of the presentation. It is poor practice to go to the end of the talk and then throw a variety of illustrations into a small period of time.

They should be clear enough to be viewed by a good-sized audience; see Chapter 19, particularly Sec. 19.7. It is not uncommon for a speaker to make the same presentation before several audiences. Therefore, aids should be prepared for presentation in a variety of physical surroundings.

Although all of us would like to be superlative speakers, most of us are not naturally so well equipped. However, a good presentation where the speaker has something worthwhile to say and says it with care and preparation can suggest a high degree of professionalism and competence.

13.1.3 General Speeches

Other types of speeches utilize the same principles. However, you will have to think of some other devices as well. For example, one of the first things you should do is to estimate what kind of audience you will have; find out as much about it as you can before you even pick your subject. Audiences can be divided into a number of groups based on age, education, social background, interests, and prejudices. You may use the same topic for any one of these groups, but if you have once given the speech before a club group you will probably find it necessary to make some important changes before delivering it to a high school audience.

Time and place will sometimes cause you to modify your speech. In general, people are more eager to learn when they are brought together during daytime hours; evening talks are more likely to be on the lighter or more inspirational side. Certain occasions seem to call for distinct types of speeches. For instance, audiences expect to be inspired and uplifted by Memorial Day talks and commencement addresses.

A cardinal principle of general speechmaking is to choose a subject that interests you and about which you know a great deal. If you are giving an oral report, this is what you will be doing anyway. But sometimes groups will try to impose a topic upon you. If it doesn't fit you, don't tackle it. If you are given free rein, choose a topic so close to you that your enthusiasm will be transmitted to your audience. There is no better way of preventing or overcoming stage fright than knowing your subject and having a tremendous interest in it. Such a combination makes you an expert; you can feel superior to your audience. And a feeling of superiority usually dissipates any nervousness you might otherwise have.

Many speeches get off to a bad start because the speaker hasn't sufficiently thought out his opening remarks. Provide a good opening for your talk by devising something that will attract attention and carry you into your specific subject. Naturally, this beginning must be compatible with your material and the audience you are addressing.

If you are giving out straight information to a class or a conference, do not cheapen your talk with a sensational beginning. On the other hand, if you are persuading the audience to accept a new idea or follow a new course of action, you may have to start with something novel or different. Stories, quotations, ironic or contradictory statements, references to well-known people—these are some of the devices that you can use.

Every speech should have a center around which everything is arranged. Ruthlessly cut out anything that does not clarify or illuminate your central ideas. Some speakers prepare too much material; they include anecdotes and stories interesting to themselves, but with little bearing on the speech as a whole. Such a speech may interest an audience momentarily, but in the long run it will only be confusing.

If your speech has a strong core, it will be much easier to put the individual points together. They may be arranged chronologically or according to importance, or built up to a series of climaxes. And the material will shape up all the better if you discard anything that doesn't really belong.

The heart of any speech is the examples used in it. It is not enough to tell your audience what the main points are. You must bring them alive. Think how effective visual aids—slides and movies—are. Since you can't provide slides and movies for every speech, do the next best thing: supply pictures and examples for the mind's eye. Every time you introduce a main topic or attempt to drive home an idea, see if you have provided a good down-to-earth example. Make it vivid and colorful. Use familiar words, action words, picture words.

The ending of a speech should be sharp and decisive. For most speeches, it is essential that you demand something of the audience, anything from merely asking it to think about a proposition to a much more positive demand like contributing money to a worthy cause. In addition to making the ending decisive, therefore, try to bring some action into it. If you have been giving information, it may be necessary to summarize what you have said. If you have been arguing some cause, come to a conclusion; if you have been talking in a light vein, end with a story.

13.2 Delivery of the Speech

13.2.1 Type of Delivery

The prepared speech (as opposed to the impromptu or spur-of-the-moment talk) can be delivered in three ways: (1) it can be read; (2) it can be memorized and recited; or (3) it can be presented extemporaneously, with or without notes. The third of these methods is usually by far the best.

The speech that is read seldom sounds fresh or vigorous; it is usually dry and humdrum. The reader's voice settles into a monotonous drone, and the listeners settle low into their seats. The reader must spend most of his time looking at his manuscript rather than at his listeners, so that he loses touch with them. As a result, he cannot tell when he needs to slow down or to provide further explanation of a point that is baffling them, or when he is boring them by dwelling too long on a point that does not interest them. The people who can read aloud in a way that is lively and stimulating are few and far between. Unless you are one of them, you will do better not to read your speeches except when you are required to. After all, what does one get from hearing a speech poorly read that cannot be gotten better by reading the paper oneself?

Very few people can sound natural or convincing when they are reciting a memorized speech. Most people have to concentrate so hard on calling up the memorized words that they lose sight of the meaning of what they are saying. Consequently their delivery is expressionless, singsong, and probably too fast. But even worse, when the reciter of a memorized speech once gets off the track, he is lost. Logic or knowledge of his subject won't tell him what to say next; only the resumption of a complicated and delicate mental pattern will get him started again, and over this process he has no control. *Don't* memorize a speech.

The extemporaneous speaker, on the other hand, has almost every advantage over the reader or the reciter. Because he is thinking what he is saying, he can speak with expression, emphasizing the important ideas. He looks at his listeners, gauges their reactions to what he is saying, and adjusts his speed and his approach accordingly. But don't get the idea that the extemporaneous speech can be dashed off with no planning or preparation beforehand. The preparation for a good extemporaneous speech requires at least as much work as the writing of a paper of the same length.

The rest of this chapter gives some pointers on the *manner* of presentation. They are aimed specifically at the extemporaneous speaker. Most of them will be helpful also to the reader (if you must be one). The reciter we will not consider at all, because we trust we have persuaded you never to memorize a speech.

13.2.2 Manner of Delivery

Some speakers have such good memories and so much confidence in themselves that they can give long and complicated talks smoothly without referring to notes. Most people, though, do better—or at least *feel* better—if they have some notes to bolster their memories. While no audience objects to a speaker's using notes, they should be as unobtrusive as possible. Don't put them on the 8½ x 11-inch paper you use for most purposes; use 3 x 5-inch cards that will fit almost unnoticeably into the palm of your hand. Make the notes just as brief as possible. Usually a bare topical outline will do, just to remind you of the things you want to talk about; sometimes you may want to jot down key phrases or important statements more fully. But don't write out your whole speech, even as a preliminary step. If you do, some of the written-language patterns may carry over into your talk.

Your manner on the platform should be dignified, yet natural. Perhaps it is idle for us to tell you to be relaxed, because you probably won't be, at least until you have had some experience at addressing gatherings, and perhaps not then. But try to be relaxed; at least try to *look* relaxed.

Speak naturally and with expression, just as you do when you are talking about, let us say, food. Avoid a monotone; modulate your voice. If you are used to talking with your hands, all right; but don't try to cultivate any special platform gestures.

Enunciate clearly. Do not run your words together; bite them off cleanly. If you are in a room of any size—say as large as a classroom or larger—you will have to speak much louder than you probably think necessary. Most beginning speakers go too fast. Speak slowly. If you have planned your speech properly, there will be no hurry.

Many speakers—including experienced ones—are plagued by the *uh* habit. Like the radio, they cannot seem to tolerate a moment of silence. But instead of filling in the gaps with foolish small talk, they use various forms of the *uh,* long or short. When you have to stop to think of what to say next, don't say *uh;* don't say *anything*. Just maintain a dignified silence, and think. Your listeners will welcome the pause; it will give them a chance to reflect—to catch up with you and consolidate their thoughts.

Always be on guard against random motions. Some speakers have little nervous habits of which they are quite unaware. Popular ones are shaking a piece of chalk as if it were a pair of dice, fiddling with the pointer, and alternately bending and unbending a knee. Any such performance is, of course, fascinating to the audience, but it tends to take their minds off what you are trying to tell them. You should not be stiff; but don't fall victim to any nervous mannerism.

Watch out also for random movement. Don't pace up and down; stay in one place. You will, of course, sometimes have to move to or from the blackboard or the screen, and these necessary moves usually provide enough exercise and variety. If you feel that a change is really necessary, take not more than one or two steps forward or backward. You want your audience to be interested in what you are saying, not where you are going.

Look at your audience. Let your eyes move slowly around the room, not neglecting the back rows. Look at your notes only as long as you need to. Be particularly careful not to let your eyes be drawn to the blackboard or the screen longer than necessary to orient yourself. Find the point on the blackboard you want to talk about, point to it, and then face your audience again, still holding the pointer to the board if necessary. Remember, you are talking to the people out front, not to the diagram on the blackboard.

The kind of blackboard presentation that develops before the eyes of the audience is often very effective, particularly for mathematical subjects. It forces the speaker to go slowly, so that his audience can keep up with him. But don't keep your audience waiting unnecessarily while you make drawings that could just as well have been put on the board ahead of time or that could be presented on slides. And use as few equations as you can.

Slides are a fine way of presenting information if they are well prepared (see Sec. 19.7). Arrange them with your talk so that you don't have to keep turning the lights on and off. That is, keep your slides in a few large groups rather than showing one here, one there. (Occasionally a whole speech is given in the dark, with frequently changed slides to accompany it all the way through.)

Just as a written document benefits by subheads, so a speech benefits by some kind of a signal when a new subject is about to come up. Sometimes a pause will do, sometimes an explicit announcement is called for: "Now I am going to tell you how the cover was welded on."

Whenever you have a serious speaking assignment, you will probably do well to practice your speech before the final presentation. Record it on a tape recorder and play it back, listening for deficiencies. Get a few friends or classmates or colleagues to act as audience and to time you. Give your speech as realistically as you can, and ask for comments and suggestions. Adjust the length to fit the allowed time. (This will practically always mean removing material from your first version.) Repeat the process until you and your trial audience are satisfied with the manner and the length of the presentation, but try to keep the repetitions to a minimum to avoid inadvertent memorization.

Finally, a word about impersonality (see Sec. 14.13). Spoken language is generally less formal than written language. As a result, the devices used to avoid the words *I* and *we* sound even more strained and artificial in a speech than they do in a report or an article. Sometimes you hear a talk that contains several references to "the speaker." Only after you have heard the phrase several times do you realize that the speaker is talking about himself, not some other speaker. Hardly any speech is such a formal occasion that you cannot speak of yourself as *I*, or of yourself and your co-workers as *we*.

III

Tools and Methods

14. Writing: Style

The principles expounded in this chapter apply particularly to technical writing. We happen to think that they apply as well to practically all other sorts of writing; but they are here directed specifically to the technical writer.

14.1 Simplicity

Remember that the primary objective of technical writing is to transmit information efficiently. The only writing style suited to this objective is simple, direct, and unadorned. Don't get the idea that an elaborate or elegant style will impress people with your erudition. For every reader impressed by bombast, ten will be annoyed or even left wondering what you are trying to say. People have neither the time nor the inclination to dig out the meaning of unnecessarily obscure prose.

In every technical reporting job, then, say things just as simply, just as directly, just as clearly as you can. Never let your sentences get involved. Don't try to impress your reader; try to tell him something.

The rest of this chapter is devoted to some specific ways of keeping your style simple, direct, and clear.

14.2 Simple Language

The first step to take toward a simple, clear style is to use simple language. Other things being equal, always choose:

(1) A short word rather than a long word.
(2) A plain, familiar word rather than a fancy, unusual word.
(3) A concrete word or a strong verb rather than an abstract word.

Lest these rules sound too categorical, notice the *other-things-being-equal* qualification. If the long word expresses your meaning better than the short word, use it. If the unusual word is more precise than the familiar word, use it. If the abstract word fits the sentence better than the concrete word, use it. But be sure, before you choose the long word or the fancy word, that your intended readers have a reasonable likelihood of knowing it.

The rest of this section consists of examples to illustrate these three rules. Each original example is a negative one—an illustration of what happens when the rules are broken. Each suggested revision shows *one way* of improving the original version. You may very well be able to think of better ways.

14.2.1 The Short Word

Original Example	*Suggested Revision*
The bumper is constructed with a steel core *encapsulated in* urethane foam.	The bumper is constructed with a steel core *covered with* urethane foam.
We expect to *commence* work on this project immediately.	We expect to *start* work on this project immediately.
The agreement was *effected*.	The agreement was *made*.
Consumption by the synthetic-rubber and explosives industries became almost negligible at the *conclusion* of 1970.	Consumption by the synthetic-rubber and explosives industries became almost negligible at the *end* of 1970.

14.2.2 The Plain (or Familiar) Word

Original Example	*Suggested Revision*
All persons working near these tubes should be *cognizant* of the danger of explosion.	*Everybody* working near these tubes should be *aware* of the danger of explosion.
The principal investigator would notify the police department *in each instance when* the medico-engineering team was ready to perform its *chronology of investigatory procedure*.	The principal investigator would notify the police department *each time* the medico-engineering team was ready to perform its *investigation routine*.

Original Example	*Suggested Revision*
This condition did not *interfere with the general adaptive capacities to provide for herself through vocational efforts.*	This condition did not *prevent her from working for a living.*
After the *completion of this methodology and protocol,* a conference was held.	*After these steps had been completed,* a conference was held.
Workers in the industry have *been desirous of improving* their wages.	Workers in the industry have *wanted to improve* their wages.

14.2.3 The Concrete Word

In general, concrete nouns name objects or things that can be perceived by the senses; abstract nouns name qualities, ideas, or conditions that are conceptions of the mind. Abstract nouns tend to be general and vague. In addition, there is a difference in force between nouns and verbs; for the most part, verbs show action and are more vital. As a result, expressions that contain abstract nouns are less forceful, less direct, less exact than their concrete counterparts, and certainly abstract nouns are much less forceful than verbs incorporating the same ideas.

Original Example	*Suggested Revision*
Fuel tank *deformation* was present.	The fuel tank was deformed.
Close *control* of temperatures, pressures, and process time is maintained.	Temperatures, pressures, and process time are closely controlled.
Uninjured *survival* of anesthetized hogs occurred in all experiments up to 80 g.	The anesthetized hogs survived uninjured in all experiments up to 80 g.
The tensometers were removed before the *advent* of large deformations.	The tensometers were removed before large deformations occurred.
There is adequate *access* to the job for men and materials.	Men and materials can reach the job readily.
Production engineers have found direct *control* of this operation to be a *necessity*.	Production engineers have found that this operation must be directly controlled.
After *setting* or *stabilization* of the mixture has taken place, the bars are removed from the molds.	After the mixture has set or stabilized, the bars are removed from the molds.
The *accuracy* of a large rotor, which gives a less intense field, is much greater.	A larger rotor, which gives a less intense field, is much more accurate.

Original Example	*Suggested Revision*
The reasons for this *popularity* are the almost universal *availability* of slide projectors and the relative *simplicity* of *preparation* of slide originals.	Slides are popular because they are easy to prepare and because slide projectors are almost universally available.

14.2.4 Short, Plain, Concrete, and Action Words

By now you have probably noticed that there is a close connection between short words, plain words, concrete words, and action words. When you choose a plain word, you are likely to get a short one; when you choose a concrete word, you are likely to get a plain one. In Secs. 14.2.1, 14.2.2, and 14.2.3 these factors are treated separately in order to make you recognize all of them. But writers who make the mistake of departing from simple language and action words usually use words that are long and unfamiliar and abstract.

The following examples do not differentiate between our three kinds of fancy language. The revised versions contain shorter, plainer, more concrete, and more action-filled words than the originals. You should find them clearer and easier to read.

Original Example	*Suggested Revision*
It was evident to the research team that when a driver is becoming familiar with a strange automobile any proclivity to collision involvement would seem to be enhanced during an adjustment period.	The research team observed that drivers seem to have an abnormal number of collisions during the period of adjustment to a new car.
Sled braking produced a smooth 2-g deceleration *until zero velocity was obtained.*	Sled braking produced a smooth 2-g deceleration until the sled stopped.
Accident frequency depends on the *adequacy* of roadway *visibility*.	Accident frequency depends on how well the road can be seen.
The initiative and enthusiasm requisite to ultimate success of the program must perforce come from the employees.	If the program is to succeed, the employees must enter into it enthusiastically.
Removal of the protecting group from the resulting phthalyl peptide *is effected* by treatment with hydrazine hydrate.	The protecting group is removed from the resulting phthalyl peptide by treatment with hydrazine hydrate.
The amplifier has been used continuously, but the reliability of operation that is desired has not been achieved.	The amplifier has been used continuously, but it has not been sufficiently reliable.

Original Example	*Suggested Revision*
The filled land has gradually compacted, and now there is no *likelihood* of further *subsidence*.	The filled land has gradually compacted, and it is not likely to sink further.
The department continued to *manifest* unusual *activity*.	The department continued to be unusually active.
One of the puzzling problems in the study of behavior pathology is that of the criteria of the norm from which the pathological manifestations are supposed to deviate.	The study of abnormal behavior is complicated by the absence of accepted standards for normal behavior.
In the coming year his assignment to the investigation is visualized.	He will probably be assigned to the investigation next year.
Measurement of the single-phase reactive component of electric energy can be *accomplished by the use of* a modified watt-hour meter.	The single-phase reactive component of electric energy can be measured with a modified watt-hour meter.

14.2.5 Dictionaries

A good dictionary is an essential tool for technical writing. It will help you choose suitable words, and it will be useful for several other purposes (see Secs. 15.10 and 16.1). By all means have one at your side and consult it frequently whenever you write.

Particularly recommended is *The American Heritage Dictionary* (Houghton Mifflin). Other dictionaries are listed in Sec. 21.3. A very useful adjunct is *Rodale's Synonym Finder* (Rodale Books), which will often help you find the word you are groping for. *Webster's Dictionary of Synonyms* (Merriam) is less valuable in supplying synonyms than in distinguishing the small differences between words with similar meanings.

14.3 Elimination of Unnecessary Words

The second step toward a simple, clear style is to eliminate from your writing every word that does not contribute to the meaning or the clarity of your message. Superfluous words are undesirable for three reasons: (1) they waste time (and paper); (2) they obscure significant words and ideas; (3) they rob statements of vigor, force, effectiveness. By all means cut them out, even though the process requires extensive rewriting.

This rule does not, of course, imply that you should write in "telegraphic" style—that you should omit the little words like articles, prepositions, and pronouns that your reader can supply from his own knowledge. You must retain these words, because they help your reader to understand you clearly and easily. They contribute to your message even though it could be deciphered without them.

Most of the unnecessary words in technical writing seem to fall into the three categories illustrated in the following three subsections.

14.3.1 Completely Superfluous Words

You will often see passages containing words that are completely superfluous—words that contribute no meaning whatever. Such deadwood should have no place in your writing. Sometimes these words can simply be struck out; sometimes their removal requires more or less recasting.

You will see that some of the examples below contain unwieldy relative clauses beginning with *which* and *that*. Others reveal trite patterns such as *in the case of*. Probably these constructions were useful at one time; now they are unnecessary.

Original Example	*Suggested Revision*
Due to the crash dynamics involved in this case whereby the brunt of the forces exerted against the vehicle were located in the front end, there was no fuel system damage.	Since this was a purely front-end collision, the fuel system suffered no damage.
I wish to *take this opportunity to* acknowledge the help of my laboratory assistant.	I wish to acknowledge the help of my laboratory assistant.
The parts *which are* in the warehouse can be shipped immediately. Those *which are* still in process will be ready in two weeks.	The parts in the warehouse can be shipped immediately. Those still in process will be ready in two weeks.
A precipitate forms in containers *that are* left undisturbed overnight.	A precipitate forms in containers left undisturbed overnight.
Significant information on vacuum-tube life has been obtained from records of tubes *which have been* used in the amplifier.	Significant information on vacuum-tube life has been obtained from the records of tubes used in the amplifier.
In *the case of* circular flight, a constant bank angle is desirable.	In circular flight, a constant bank angle is desirable.
Measurements were also taken with a second probe at the center of the discharge. *In this case* the probe was surrounded by a guard ring as before.	Measurements were also taken with a second probe at the center of the discharge. The probe was surrounded by a guard ring as before.
Only seals *which are of the* hermetic *type* give complete protection.	Only hermetic seals give complete protection.
A crystal rectifier has been connected across the field to reduce *the magnitude* of the voltage.	A crystal rectifier has been connected across the field to reduce the voltage.

Original Example	*Suggested Revision*
A 10-inch reservoir *is placed* between the main stopcock and the oil diffusion pump. *This* permits operation of the diffusion pump *to be* continued in the event that a short-term shutdown of the forepump *becomes necessary*.	A 10-inch reservoir between the main stopcock and the oil diffusion pump permits continued operation of the diffusion pump in the event of a short-term shutdown of the forepump.
This knowledge is of particular importance in the operation of a system like the one described in this report. *This can be realized when attention is directed to the fact that* a single error can render such a system worthless.	This knowledge is of particular importance in the operation of a system like the one described in this report, *because* a single error can make such a system worthless.
Figures 6 and 7 show oscillograms of the switching operations, and *it can be seen* that the error is only a few microseconds.	Oscillograms of the switching operations, Figures 6 and 7, show that the error is only a few microseconds.

14.3.2 Circumlocutions

A circumlocution is a roundabout expression—an expression that uses several words to express an idea that can be expressed in fewer words or in one word. Occasionally a circumlocution is desirable: if, for example, your audience is not likely to know its single-word equivalent. (Should we have said *roundabout expression* instead of *circumlocution?*) But usually circumlocutions just get in the way. Try to avoid them.

Original Example	*Suggested Revision*
This analysis implies reduced failure rates for the current year, but the data are not sufficiently complete to *allow the conclusion that such is an actual fact*.	This analysis implies reduced failure rates for the current year, but the data are not sufficiently complete to *be conclusive*.
The control panels are *overly crowded to the extent* that it would require hours to replace some of the components.	The control panels are *so crowded* that it would require hours to replace some of the components.
For this reason, a steam plant is the Company's main source of power.	*Therefore*, a steam plant is the Company's main source of power.
Independent suspension of the front wheels is desirable *due to the fact that* it permits softer front springs to be used.	Independent suspension of the front wheels is desirable *because* it permits softer front springs to be used.
Despite the fact that wages are high in the textile industry, unions are strong.	*Although* wages are high in the textile industry, unions are strong.

Original Example	*Suggested Revision*
To *make an approximation as to how much* time it might take to establish deflection, the following equation is useful.	To *estimate* the time it might take to establish deflection, the following equation is useful.
It is in recognition of this fact that our vacuum-system bores are as large as other practical considerations permit.	*Therefore* our vacuum-system bores are as large as other practical considerations permit.
Leaving out of consideration the size of the tube sealed to the system . . .	*Disregarding* the size of the tube sealed to the system . . .
The tensile strength of gray cast iron differs from the compressive strength *by an extremely large amount.*	The tensile strength of gray cast iron differs *markedly* from the compressive strength.
The nature of the product *is such as to fully negate the feasibility of* an automatic process.	The nature of the product *precludes* an automatic process.
Although *the greatest percent* of the cameras are sold outside of the industry . . .	Although *most* of the cameras are sold outside of the industry . . .

14.3.3 Indirect Expressions

Under this heading we have grouped several constructions that have two characteristics in common: (1) they say things indirectly and weakly rather than directly and vigorously; (2) they use more words than their direct equivalents.

When you look at the examples, you will see that many of the indirect expressions begin with pronouns followed by weak verbs. Such constructions throw sentences out of balance, leaving the emphatic parts until the end. Begin your sentences as vigorously as possible.

Original Example	*Suggested Revision*
There are three basic requirements *that* a saw-tooth sweep signal must fulfill. *These are:* (1) . . . , (2) . . . , (3). . . .	A saw-tooth sweep signal must fulfill three basic requirements: (1) . . . , (2) . . . , (3). . . .
There are approximately 7000 persons *who* reach their tenth birthday daily in the United States.	Every day about 7000 children in the United States reach their tenth birthday.
There is a city-owned pier running out from this land *which is* used by a marine-repair firm.	A city-owned pier running out from this land is used by a marine-repair firm.

Original Example	*Suggested Revision*
It was probably because of this increased demand *that* the price of cadmium rose sharply.	Probably because of this increased demand, the price of cadmium rose sharply.
It might be expected *that there would be* some interference to the television channels.	Some interference to the television channels might be expected.
It is essential that the requirements of all the control systems with which the switch may be used be taken into consideration.	The requirements of all the control systems with which the switch may be used *must* be taken into consideration.
It is felt that the derivations of these equations will be interesting to those who have been working on the problem.	The derivations of these equations *should* be interesting to those who have been working on the problem.
It appears that the synthetic material is better than the natural.	The synthetic material is *apparently* better than the natural.
It should be noted that the amount of brightwork was increased despite shortages of copper and nickel.	*Note* that the amount of brightwork was increased despite shortages of copper and nickel.

Is the imperative too blunt? Probably not; but you might have to soften it in some situations.

Original Example	*Suggested Revision*
It is obvious that the city should increase its reserve water supply.	*Obviously* the city should increase its reserve water supply.
It is found empirically that the disintegration of a radioactive element *follows a simple decay law which states that* the rate of decay is proportional to the amount of the radioactive element present.	It is found empirically that the rate of disintegration of a radioactive element is proportional to the amount of the radioactive element present.
In the case of radiation studies *it was found that* . . .	Radiation studies *have demonstrated that* . . .
By looking at the bar graph on page 20, *one can see that* . . .	The bar graph on page 20 shows that . . .
In a triangle with three equal sides *it may be shown that* the three angles are also equal.	In a triangle with three equal sides the three angles are also equal.

It may be shown that is a phrase much used by authors of textbooks who wish to save space by omitting proofs or derivations. It is also popular with students who have forgotten proofs or derivations. The phrase may often be omitted. If the reader is familiar with the proof, he does not need to be told it exists; if he is not familiar with it, he is not enlightened by the mere statement that it exists. Sometimes *it may be shown that* can be replaced by a more informative or more useful statement, as in the next two examples.

Original Example	*Suggested Revision*
It may be shown that $\frac{V_1}{T_1}=\frac{V_2}{T_2}$.	According to Charles' law, $\frac{V_1}{T_1}=\frac{V_2}{T_2}$.
If these two currents are direct currents, it may be shown that their sum can have only two possible numerical values.	If these two currents are direct currents, by Kirchhoff's first law their sum can have only two possible numerical values.

Some particularly weak indirect statements that are often used in an overzealous effort to maintain impersonality are illustrated in Sec. 14.13.

14.4 Involved Sentences

Some sentences are difficult to read and understand because they are *involved:* their flow of thought is not straightforward. Involved writing may result from the elementary error of sentence elements in the wrong order (see Sec. 15.3); long sentences are more likely to be involved than short ones (see Secs. 14.5 and 14.6). But short sentences can be involved too, and often involvement is a generally muddy quality that is hard to analyze. One kind of involved sentence in particular that seems to plague the technical writer is the sentence containing a large number of prepositions, especially the preposition *of.*

When you are reading over your rough drafts, watch out for sentences like these.

Original Example	*Suggested Revision*
Often the ratio of stress concentration here to that elsewhere runs as high as or higher than 3:1.	Often the stress concentration at these points runs 3 times or more as high as it does elsewhere in the specimen.
Above 450 C, the silver content was found not to be constant as was expected.	Contrary to expectations, the silver content was found not to be constant above 450 C.
A useful method is that of Evans and Goodman, which involves the volumetric determination of helium in a rock sample and measurement of the decay constant by measuring the rate	A useful method is that of Evans and Goodman. In this method the volume of helium in a rock sample is determined; then the decay constant is determined by measurement of the

Original Example	*Suggested Revision*
of production of alpha particles, which are the only source of helium in rocks.	rate of production of alpha particles (the only source of helium in rocks).
Thus the importance of composition is shown if it is assumed that the other variables which may affect the quality of a sinter were held constant and were of such a value as to produce, in conjunction with the proper composition, a superior sinter.	If it is assumed that the other variables were proper and constant, this test demonstrated the importance of composition to the quality of a sinter.
The properties of electromagnetic propagation in a long wire of a transmission system is the subject of this paper.	This paper discusses the properties of electromagnetic propagation in transmission systems containing a long circular wire.
The roasting technique was not as effective as was expected and needs further investigation in order that the factors which prevented the attainment of the quality of roast desired may be determined.	The roasting operation left more sulfur in the sample than was expected. The technique needs further investigation to determine the factors that prevented a satisfactory roast.

14.5 Supercharged Exposition

We have used the adjective *supercharged* to describe exposition that has been pumped so full of meaning that the reader has difficulty in containing it and conquering it. The next two subsections illustrate two kinds of supercharged exposition. The second of these is more frequent and more dangerous than the first.

14.5.1 Too Much Pruning

In a worthy zeal to eliminate words (see Sec. 14.3), the technical writer sometimes boils down a statement until it contains insufficient information. The remedy is to complete the explanation; or perhaps occasionally to add words just to dilute a passage that is too concentrated for human consumption.

Original Example	*Suggested Revision*
The electronic circuit requires a switching time of 1 millisecond. This time, which is due to transients acting on the amplifier, would also be present in a relay system; therefore, a time saving of the action time of the relays has been achieved.	The electronic circuit requires a switching time of 1 millisecond. This time is the time required for decay of the transients acting on the amplifier. The same transient-decay time would be required in a relay system; therefore the electronic control results in a net saving of the time required for operation of the relay, or about 10 milliseconds.

Original Example	*Suggested Revision*
On the west the watershed is more centrally located rather than near the southern coast as in the east.	On the west *side of the island* the watershed is more centrally located, rather than *being* near the southern coast as *it is* on the east *side*.
Progress is being made *to show* no significant difference between the variability from day to day and within the day.	Progress is being made *toward a demonstration that there is* no significant difference between the variability from day to day and *the variability* within the day.
The results are shown in Fig. 3. The curves point up the great increase in the maximum conductivity and the gradual shifting of the maximum in the direction of the more concentrated solutions.	Fig. 3 presents a family of curves of conductivity vs. concentration at four different temperatures. Notice that there is a maximum conductivity for each temperature. Also, as the temperatures increase, the conductivity curves are higher and the maximum is displaced in the direction of the more concentrated solutions.

14.5.2 The Long Sentence

A long sentence that flows smooth and straight—a sentence that is not involved—can still contain so much information that the reader's mind can't keep up with it and can't assimilate its information. The remedy for this kind of supercharged exposition is to break the long sentences up into pieces of more easily digestible size. This process often requires the addition of a few accessory words.

Original Example	*Suggested Revision*
Both vehicles were proceeding south on Post Road, and both were driving straight ahead up to the point where the struck vehicle was attempting to make a right turn, rather slowly, into a driveway west of Post Road when impacted by the Pontiac station wagon.	Both vehicles were proceeding straight south on Post Road. Then the struck vehicle attempted to make a right turn, rather slowly, into a driveway to the west. At this point it was impacted by the Pontiac station wagon.
The fixed-point logic requires that care be taken in planning the computation to insure that all numbers be less than unity in magnitude, consequently requiring that the upper bounds of the magnitudes of all variables be known at least approximately and that quantities which would normally be greater than one be premultiplied by some number,	The fixed-point logic requires that care be taken in planning the computation to insure that all numbers be less than unity in magnitude. Therefore (1) the upper bounds of the magnitudes of all variables must be known at least approximately, and (2) quantities that would normally be greater than one must be premultiplied by some number

Original Example	*Suggested Revision*
usually a negative power of two, to make them smaller than one.	(usually a negative power of two) to make them smaller than one.
Suppose a collection of vectors is such that the sum of any two vectors of the collection also belongs to the collection, and such that the scalar product of any vector of the collection with any real number is a vector belonging to the collection.	Suppose that a collection of vectors satisfies two conditions: (1) the sum of any two of the vectors is also a member of the collection; (2) the scalar product of any vector of the collection and any real number is a vector belonging to the collection.
If we recall that the points of a plane can be represented by a couple of numbers (x, y), where x is the distance of the point to the right of an arbitrary origin and y is the distance above that origin, we can represent vectors in the plane as a couple of numbers (x, y), provided we put the rear ends of the vectors at the origin, since vectors may be moved in space unchanged so long as their direction is not altered.	Let us recall two propositions: (1) Any point of a plane can be represented by a couple of numbers, x, y (x is the distance of the point to the right of an origin, y the distance above that origin). (2) Vectors may be moved in space unchanged so long as their direction is not altered. Therefore, if we put its rear end at the origin, we can represent any vector in the plane by a couple of numbers, x, y.

14.6 Varied Sentence Length

You have seen in Secs. 14.4 and 14.5 that long sentences can be very troublesome to the reader if they are involved or if they contain too much high-powered information. Indeed, a current school of thought holds that any prose with an average sentence length of more than 17 words is difficult for the average reader to understand. But for two reasons you should probably vary the length of your sentences, mixing in a few long ones with the short ones: (1) Some of the concepts you will write about are too big or too complex to be expressed in very short sentences. (2) An uninterrupted succession of short, declarative sentences produces a very unpleasant and monotonous effect.

Caution: If you are unable to write long sentences that are clear and uninvolved, by all means stick to short ones. It is better to be clear and jerky than obscure and smooth. And don't overdo the admixture of long sentences.

Here are some examples of jerky, short sentences that can easily be combined and smoothed out with no serious increase in difficulty.

Original Example	*Suggested Revision*
An adjustment box provides for calibration of the instrument. The box may be located in the cockpit or on the instrument panel.	An adjustment box in the cockpit or on the instrument panel provides for calibration of the instrument.

Original Example	*Suggested Revision*
It is a single-family dwelling of two stories and has an adjoining garage. The house is 36 feet long and 24 feet wide. The joists are lapped and sit on top of the sills, girders, and plates. The joists are bridged with 1 x 3 bridging every 6 feet. The roof is a hip roof and has a slope of 9/12. All sheathing is 10-inch square-edged boards. Many ¼-inch cracks were left between the boards. The firestopping between all joists is brick laid in mortar. It extends 5 inches up into the walls.	The house is a single-family dwelling of two stories, 36 feet long and 24 feet wide, with an adjoining garage. The joists, which are lapped, sit on top of the sills, girders, and plates, and are bridged with 1 x 3 bridging every 6 feet. The roof is a hip roof with a slope of 9/12. All sheathing is 10-inch square-edged boards, with many ¼-inch cracks between them. The firestopping between all joists—brick laid in mortar—extends 5 inches up into the walls.
The coke available was too large to use with good results. The crews broke the coke into usable pieces with hammers. Different particle sizes in the charges were tried. The chemical analysis of all the cokes was the same. The ash content was 11 percent.	Since the coke available was too large to use with good results, the crews broke it into usable pieces with hammers. Different particle sizes in the charges were tried. The chemical analysis of all the cokes was the same, and the ash content was 11 percent.

14.7 The Passive

Every textbook on composition stresses the idea that the passive voice is less vigorous and less effective than the active. Its use is likely to result in wordy, indirect, vague passages. Yet in technical writing the passive inevitably appears a great deal, because of the requirement for impersonality (a requirement that is much exaggerated; see Sec. 14.13). As a result, many technical writers get into the habit of using the passive much more than they need to. Your writing will be clearer and more forceful if you say everything you can in the active rather than the passive.

The choice between active and passive may sometimes be based on what is being stressed. If it is the *agent,* the active is appropriate; if the *act,* the passive.

Here are some examples of conventional passive statements and their active equivalents. In the first one, notice the important gaps in the information transmitted. *Who* appointed the committee? *Who* discussed the possibility of a joint program? *Who* made the request? *Who* is going to release results?

Original Example	*Suggested Revision*
A committee was appointed to coordinate efforts of the Society of Automotive Engineers and the National	The Society of Automotive Engineers appointed a committee to coordinate their efforts with those of the Na-

Original Example	*Suggested Revision*
Highway Safety Bureau. The possibility of a jointly sponsored research program was discussed. It was requested that all decisions concerning a joint program be deferred until the results of two existing research contracts have been released.	tional Highway Safety Bureau. The committee discussed with representatives of the NHSB the possibility of a jointly sponsored research program. The NHSB requested that all decisions concerning a joint program be deferred until the Bureau had released the results of two existing research contracts.
In the last report the importance of maintaining a uniform spacing between adjacent cylinders was described.	The last report described the importance of maintaining a uniform spacing between adjacent cylinders.
Oxidation tendency is also accelerated by the presence of metals that act as catalysts.	The presence of metals that act as catalysts also accelerates oxidation.
Reports have been presented by several members.	Several members have presented reports.
The protecting group is removed from the resulting phthalyl peptide by treatment with hydrazine hydrate.	Treatment with hydrazine hydrate removes the protecting group from the resulting phthalyl peptide.

For examples of some very weak passives used in order to achieve impersonality, see Sec. 14.13.

14.8 Technical Jargon or Shoptalk

The term *jargon* is used to describe two different kinds of language, one of them much more objectionable than the other. In its broad sense, jargon is any loose, fuzzy, unintelligible talk or writing. For an amusing, eloquent, and wordy plea against this kind of jargon, see the chapter "On Jargon" in Sir Arthur Quiller-Couch's *On the Art of Writing* (Sec. 21.3).

In a narrower sense, jargon is the specialized technical vocabulary peculiar to any trade, science, profession, or special group of people. This kind of jargon—it might better be called *shoptalk*—is often condemned wholesale, along with the other kind. The main drawback of shoptalk is that it is a "secret" language: an engineer will not understand a physician's shoptalk any more than a physician will understand a bricklayer's. Indeed, it is this very secrecy that makes shoptalk appeal to some people (think of the conversation of some radio amateurs).

But shoptalk very often expresses in one word a large concept that would require many words, or even sentences, from the general vocabulary. For instance, when the computer engineer speaks of *hardware,* his brothers in the trade know that he is talking about the physical units—the electronic, mechanical, and other gear—at the heart of an information-processing system. When he speaks of *software,* they know that he is talking about the programming functions—the paperwork, flow charts, punched cards, magnetic tapes, magnetic disks, and other materials used in feeding information to the computer—and the output, in the form of plots, printed tables, or whatever.

If he is talking to his colleagues, this computer engineer would be foolish to spend the time and words necessary to avoid the use of *hardware* and *software.* On the other hand, if he is addressing laymen, he would be wasting his time to speak of hardware and software, because they might not know what he was talking about.

This discussion of shoptalk can be generalized into two rules:

(1) When a term from the general vocabulary expresses your idea as well and as economically as a shoptalk term, always choose the general term.

(2) Even if a shoptalk term is more precise or more economical than a general term, do not use it unless you are sure that your audience will know it. (When you are going to express the idea repeatedly, you can use the strange term if you define it carefully at its first appearance.)

Of course slangy shoptalk is as inappropriate as any other slang in formal situations.

Below are several examples of shoptalk that could easily be expressed in terms from the general vocabulary.

Original Example	*Suggested Revision*
Analog data were recorded on magnetic tape and then digitized by an A/D converter.	Analog data were recorded on magnetic tape and then converted to digital form by an analog/digital converter.
One further restraint on the system design process: we should not suboptimize.	. . . : we should not overoptimize any subsystem. *or* . . . : we should not improve any one part of the system to the point where the performance of other parts is worsened through overload or neglect.
The lesion was discovered through an IVP.	The lesion was discovered through an intravenous pyelogram. *or* The lesion was discovered through an x-ray study of the pelvic region with an opaque dye introduced into the veins.
When such a system is studied *in vitro,* it demonstrates respiratory activity at a rate proportional to the concen-	When such a system is studied *in the test tube,* it demonstrates respiratory activity at a rate proportional to the

Original Example	*Suggested Revision*
trations of these breakdown products. *In vivo,* the concentration is determined by the rate of work deing done by the cell.	concentrations of these breakdown products. *In the living cell,* the concentration is determined by the rate of work being done by the cell.
The new engine *peaks* at 5500 rpm.	The new engine *produces its maximum horsepower* at 5500 rpm.

Several years ago Arthur D. Little, Inc., a consulting engineering firm of Cambridge, Massachusetts, published an issue of their *Technical Bulletin* called "The Turbo-Encabulator," which received wide circulation and was even reprinted in a popular weekly magazine. Here is the opening paragraph:

> For a number of years now work has been proceeding in order to bring to perfection the crudely conceived idea of a machine that would not only supply inverse reactive current for use in unilateral phase detractors, but would also be capable of automatically synchronizing cardinal grammeters. Such a machine is the "Turbo-Encabulator." Basically, the only new principle involved is that instead of power being generated by the relative motion of conductors and fluxes, it is produced by the modial interaction of magnetoreluctance and capacitive directance.

"The Turbo-Encabulator" is the *reductio ad absurdum* of technical jargon. It is full of specially composed, trumped-up, absurd technical terms that mean nothing at all. But it sounds so much like many real engineering reports that most laymen could read it without ever finding out that it is a hoax, and many engineers get halfway through it before they discover that their legs are being pulled. Don't let your writing get into the "Turbo-Encabulator" class.

14.9 Clichés

A cliché is a trite, overused expression or combination of words. The cliché is usually a *tricky* phrase that was fresh and vigorous when it was first coined because it involved a certain surprise element. But when such a phrase is used over and over again, it becomes so familiar that it loses all its impact. Like an overplayed phonograph record, it is hardly heard.

Have you ever noticed that a gaudily decorated nightspot looks much worse when old and worn than does an unpretentious tool shed? Similarly the overused cliché is more noticeable and more unpleasant than the often-repeated everyday expression. Try to avoid figures of speech that have become common. Here are a few examples.

Original Example	*Suggested Revision*
He *left no stone unturned* in his efforts to achieve tenure. Finally, *a sadder but a wiser man,* he learned that *in this day and age,* tenured professorships are *few and far between.* His campaign *ground to a halt,* and at subsequent faculty meetings he was *conspicuous by his absence.* He concluded his farewell to his students with these *words of wisdom: "Last but not least,* follow this advice: *do as I say, not as I do."*	He tried strenuously to gain tenure. But finally he became aware that few tenured professorships are available. He ceased his efforts, and he stopped attending faculty meetings. In his farewell to his students, he exhorted them not to follow his career as an example.
For any engineer *with his ear to the ground,* Dr. Smith's paper on atomic energy furnishes *food for thought.*	Any engineer alert to the problem of dwindling fuel supplies would find Dr. Smith's paper on atomic energy stimulating and provocative.
Must be seen to be appreciated.	[You better think up something entirely different for this old chestnut.]

Let us temper this advice, though. A few well-worn expressions that border on being considered clichés are so graphic and so economical that to avoid them would be futile. For example, the expression ". . . can't see the forest for the trees" communicates a large idea so clearly and so succinctly that any conceivable substitute is likely to be inferior. Similarly, the familiar phrases "practice what you preach," "hit the nail on the head," and "wear and tear" are hard to improve on. You will find othei old but felicitous expressions that are still vigorous enough to be worth using.

14.10 Repetition

At some time during your schooling, you were probably told not to repeat a word within a sentence, or within a paragraph, or within some other unit. This is dangerous advice; you should not take it literally. *It is always better to repeat a word than to becloud your meaning or to sound artificial in avoiding the repetition.* The *contrived variation* can be very confusing:

Original Example	*Comment*	*Suggested Revision*
Circuit theory and field theory are complementary disciplines, each very useful. Of the many engineering applications of stereonics, the following may be mentioned.	Even when you have recently been told that *stereonics* is another name for *field theory,* this passage at first reading seems to mention three branches of learning rather than two.	Circuit theory and field theory are complementary disciplines, each very useful. Of the many engineering applications of field theory, the following may be mentioned.

Original Example	*Comment*	*Suggested Revision*
It is best to start with a coarse net, even though a fine mesh is the ultimate goal. Successive trials bring the grid lines closer and closer together.	The effort required to correlate *net, mesh,* and *grid* in this passage delays the reader's understanding.	It is best to start with a coarse grid, even though a fine one is the ultimate goal. Successive trials bring the grid lines closer and closer together.
In this collection of vectors, the sum of any two vectors is a member of the set.	Did you realize immediately that *set* was the same as *collection?*	In this collection of vectors, the sum of any two vectors is a member of the collection.
Progress has been defined as the replacement of an old set of problems by a series of entirely new ones.	Did the author mean to differentiate between *set* and *series?* Probably not; but the variation sets us to wondering, and it sounds contrived and artificial.	Progress has been defined as the replacement of an old set of problems by a set of entirely new ones.

It is the repetition of hackneyed phrases and big or unusual words that is objectionable, not the everyday bread-and-butter words. You eat bread and butter with every meal and don't get tired of it; but try that with plum pudding! Consider this excerpt from a paper on the Port of Baltimore:

> Each of the trunk railroads wants to obtain as much traffic as it can with independence from the other carriers, but a certain interchange of traffic is necessary. It is believed that the existing lighterage system allows this interchange with the least capital expenditure.
>
> It is believed that there are four pieces of land in the port having possibilities for development as ship terminals.
>
> The Hawkins Point land is owned by the Patapsco Land Co. Being adjacent to the Davison Chemical Co. plant, it has railroad connections. It is primarily open land and is within the lighterage limits of the harbor.
>
> It is believed that this section is not entirely satisfactory, for the following reasons:
>
> 1. It is too far from the business center of town and the existing piers and warehouses.
> 2. It is served by only one railroad.
> 3. It is believed that for the best interests of the industrial development of the city, outlying land of this type should be used by industry.

You probably noticed that *It is believed that* occurs four times in this passage, and you were probably annoyed by it. (The phrase could be eliminated all four times: the first statement might then be softened by "The . . . system *seems* to allow . . ."; the second by "Four pieces of land . . . *show* possibilities. . . .") But did you notice that the word *land* occurs five times? And that *traffic* appears twice in one sentence, and the businesslike *interchange* in successive sentences?

The repetition of even a short, everyday word can become monotonous if it is overdone:

Original Example

Although cadmium has been known as a metal since 1817, practically it is one of the younger metals because its use has not been fully explored and developed. Until the early 1920's there were only a few uses for cadmium, and its occurrence in ores was looked on solely as a nuisance. When it was discovered that the metal could be recovered as a by-product in the electrolytic zinc process, no use was known for it, and early attempts to substitute it for tin were unsuccessful. Later its valuable properties in rust-proof coatings on steel became known, and then followed its use in high-pressure bearing alloys. More uses were found for the metal, and its importance increased rapidly; in 1941, the consumption of cadmium reached nearly 4000 tons in the United States alone. Even then the demand was so great that it was placed on the list of critical materials for war uses, and its use was restricted by the War Production Board.

Well over half of the cadmium produced is used for the electroplating of other metals. The aircraft and automotive industries, which are the principal consumers, use a large number of cadmium-plated bolts and parts.

Suggested Revision

Although cadmium has been known as a metal since 1817, practically it is one of the younger metals because its applications have not been fully explored and developed. Until the early 1920's there were only a few uses for cadmium, and its occurrence in ores was looked on solely as a nuisance. When it was discovered that the metal could be recovered as a by-product in the electrolytic zinc process, no use was known for it, and early attempts to substitute it for tin were unsuccessful. Later its valuable properties in rust-proof coatings on steel became known, and then followed its use in high-pressure bearing alloys. More applications were found for the metal, and its importance increased rapidly; in 1941, the consumption of cadmium reached nearly 4000 tons in the United States alone. Even then the demand was so great that it was placed on the list of critical war materials, and its consumption was restricted by the War Production Board.

Well over half the cadmium produced goes into the electroplating of other metals. The aircraft and automotive industries, which are the principal consumers, employ a large number of cadmium-plated bolts and parts.

The word *use* appears nine times in the original example, three times in the revision. Yet the variations do not seem contrived.

Repetition is particularly objectionable when the repeated word stands for two different ideas. This kind of repetition is not just monotonous; it is confusing.

Original Example	*Suggested Revision*
Smith went to some *pains* to make the revision as *painless* as possible.	Smith went to some *trouble* to make the revision as *painless* as possible.
Fuel density *varies* with temperature. Although there is some *variation* in different grades of gasoline, the value of 0.1 percent per degree centigrade is typical.	Fuel density *varies* with temperature. Although the *variation* is not the same for all grades of gasoline, the value of 0.1 percent per degree centigrade is typical.

Varies refers to density. The original *variation* refers to differences between fuels. *Variation* in the revised version refers, again, to density.

In summary: try to avoid monotonous repetition; but always use the best, the clearest word you can find, repetition or no repetition.

14.11 Footnotes (Explanatory)

You may wonder why a discussion of footnotes (the further-explanation kind) appears under the heading of "Writing: Style" rather than under "Mechanics," along with the treatment of citation or reference footnotes in Sec. 17.6. We have chosen to talk about explanatory footnotes here because they have a pronounced effect on style and general tone and clarity.

Every footnote is a digression from the main stream of thought. It interrupts the reader. If he reads it, his eyes must make an excursion to the bottom of the page (or to the end of the chapter); and when he is through, he must find the place where he departed from the main text. Lengthy footnotes can cause the reader to lose the thread of your message, to miss what you are trying to tell him. Even short footnotes are at best a nuisance. Use as few of them as you can.

When we borrow the words or the ideas of somebody else, we must give credit for them. Reference footnotes furnish a convenient and unobtrusive way of fulfilling this obligation. Most people do not bother to read them except occasionally.

But explanatory footnotes can practically never be justified on the grounds of necessity. Most of them can be blamed on one of two factors: (1) the writer has the mistaken notion that a lot of footnotes will make his discourse look learned or scholarly or elegant; (2) the writer has not taken the trouble to organize his material thoroughly, and he finds that he has some good ideas that do not fit readily into his main text.

We believe that practically all the customary further-explanation footnotes should either be integrated into the main text or relegated to the wastebasket (or perhaps put in an appendix). Whenever you find yourself with some leftover comments that you are tempted to put into a footnote, evaluate them carefully. If they really contribute to your theme, take the trouble to work them into the text (perhaps in parentheses, if they are of secondary importance). If they are not worth this much trouble, don't bother your reader with them at all.

Here is a brief example of a footnote that can easily be moved to the text, where it is much less of an interruption.

Original Example	*Suggested Revision*
The resultant voltage induced around a loop * by an incident wave is found as follows. Voltages induced in the vertical members are equal to . . .	The resultant voltage induced around a loop by an incident wave is found as follows. (A square loop is assumed. A round loop would be reduced to an equivalent square.) Voltages induced in the vertical members are equal to . . .

Particularly annoying is an undifferentiated mixture of citation footnotes and explanatory footnotes. Confronted by such a mixture, the reader who is in the habit of ignoring citation footnotes is likely to miss essential information in the explanatory footnotes; or the conscientious reader who doesn't want to skip any worthwhile information finds himself being continually interrupted by purely routine citations.

14.12 Subordination

A common failing of technical writers is to express ideas of unequal importance in constructions that seem to give them equal weight. Your readers will grasp your meaning more quickly and more easily if you indicate subordinate ideas by putting them in subordinating constructions. At the same time you will probably eliminate some unnecessary words, and you will relieve your style from the monotony of a succession of simple declarative sentences.

In the first example, the more important idea is that the figures represent the energy consumed each year; the fact that the figures are in kilowatt-hours should be subordinated to the main idea and take second place. A good general rule is that whenever you write a sentence with the ideas joined together by *and* or *so,* see if it wouldn't be improved by subordination.

Original Example	*Suggested Revision*
These figures are in kilowatt-hours and represent the energy consumed each year.	These figures (in kilowatt-hours) represent the energy consumed each year.
This value is best determined by actual test and is 50 watts.	This value, best determined by actual test, is 50 watts.
Sand is the other important raw material and is procured from an outside supplier.	Sand, the other important raw material, is procured from an outside supplier.

* A square loop is assumed. A round loop would be reduced to an equivalent square.

Original Example	*Suggested Revision*
The maximum obtainable temperature was found to be 1800 F and was maintained throughout ⅔ the length of the tube.	The maximum obtainable temperature, which was found to be 1800 F, was maintained throughout ⅔ the length of the tube.
This estimate has been plotted in Fig. 3 and shows the likelihood that the meters will all fail at the same time.	This estimate, which has been plotted in Fig. 3, shows the likelihood that the meters will all fail at the same time.
Applied Science and Technology Index is published in New York and it lists periodical articles on technology and commerce.	*Applied Science and Technology Index,* published in New York, lists periodical articles on technology and commerce.
The cornice detail is simple and is illustrated in Fig. 1.	The cornice detail (illustrated in Fig. 1) is simple. *or:* The simple cornice detail is illustrated in Fig. 1.

14.13 Impersonality

The last section in this chapter on style is probably one of the most important and certainly the most controversial.

Most technical reports and papers describe ideas or physical events or physical objects. Quite properly, the observations are supposed to be accurate and objective, and the observer is thrust into the background. When you report on an investigation, your reader is interested in what you found out, not in you. So far, so good.

Based on this perfectly sound notion a convention has grown up that the technical writer must always remain completely impersonal; in particular, that he must never use the first-person singular pronoun, *I*. As a result, most technical reporting is much more impersonal and much drier than it really needs to be. Don't you increase the efficiency of a communication when you make it interesting and pleasant to read? Contrast these two excerpts from departmental reports:

The Impersonal Approach	*The Personal Approach*
The staff in these trying times, although pressed by routine duties, has nevertheless found time to participate in professional and scholarly activities. Meetings of such professional societies as the American Association for the Advancement of Science and the American Society for Engineering Education have been well attended. In	In the spring of 1970 the Department agreed to make an investigation and report for the government on the development of waterpower facilities. This task enlisted the services of over 50 professional men, including mechanical, civil, and electrical engineers, physicists, and meteorologists. As director of the project, I had unusual opportu-

The Impersonal Approach	*The Personal Approach*
some instances papers have been read, and an appreciable quota of committee appointments have been assigned to Department staff members. The Chairman of the Department has been selected as the editor of a forthcoming symposium of studies relating the various branches of engineering to the general advancement of society.	nity to observe the benefits derived from an enterprise which united outstanding men from several universities and from industry—men with widely differing professional backgrounds but with a common goal.

These excerpts come from reports written and signed by the respective department chairmen. The one in the left column uses the conventional, proper, impersonal approach. It is rather stuffy; you wonder for a moment just who has been attending meetings and reading papers. Is the chairman really being impersonal and modest when he speaks of himself as *the Chairman of the Department?*

The report in the right column relaxes a little from the purely impersonal, third-person approach. To us it is a very refreshing change from the more conventional style. Does the chairman sound egotistical, or does he lose his objective viewpoint, when he speaks of himself as *I?* We think not.

Many technical reports and papers describe work that is so objective that the reader does not care who did it. He wants to know that the work was done, but he is not at all interested in the agency that brought it about. The conventional impersonal passive construction (see Sec. 14.7) is suitable for this kind of subject.

But often—although you are completely objective and suitably modest—you need to introduce the concept of yourself in order to describe the work that you have done or to warn your readers that you are expressing an opinion. If you are slavish about avoiding *I,* you have in general three choices:

(1) You can call yourself *the author,* or *the writer,* or *one,* and talk about yourself in the third person. Now we believe—and a lot of people agree with us—that this construction is highly artificial and that it is in fact no more impersonal than *I.* Furthermore, it is likely to be confusing, because "the author" could be talking about some other author.

(2) You can resort to an extremely weak passive construction in which the doer is unnamed and vague, although his identity is really an essential part of the story.

Original Example	*Suggested Revision*
The first week of the period was spent on vacation. Work is now continuing on completion of the next annual report. A description of the new high-temperature technique is the item that has come under consideration.	During the first week of the period I was on vacation. I am now writing a description of the new high-temperature technique for the next annual report.

Original Example	*Suggested Revision*
The current period has been devoted to becoming familiarized with the new equipment.	I have spent the current period in becoming familiarized with the new equipment.

These examples illustrate perhaps the most absurd place to be literal about impersonality—the progress report on your own activities. You are talking about yourself, and you may as well admit it.

(3) You can use the weak indirect construction discussed in Sec. 14.3.3.

Original Example	*Suggested Revision*
It is believed that the city should increase its reserve water supply.	I believe that the city should increase its reserve water supply.
It is desired to ascertain how you succeeded in increasing the yield of low-boiling fractions.	We should like to find out how you increased the yield of low-boiling fractions.
It was a pleasure to talk with you and your staff.	I enjoyed talking with you and your staff.
It is my thinking (thought) that . . .	I think (or believe) that . . .

All three of these alternatives are at best artificial, stilted, and vague; at worst, downright confusing. There is only one natural, direct way to talk about yourself, and that is to say *I*.

Similarly, the natural word to use when you are speaking for a group of people is *we*. Authors of textbooks use *we* very freely to stand for an indefinite body of people that includes the writer and his readers. But *we* meaning the writer and his associates or colleagues is treated with the same prejudice as *I*.

You can sound egotistical and subjective without ever uttering an *I* or a *we*. Conversely, you can maintain the proper tone of modesty and objectivity even though you do say these words. If you will use the first-person construction judiciously, you will often find your writing task eased, because you will be speaking naturally; and your audience will find the result clear and pleasant to read.

For some other words on the use of the personal pronouns in technical writing, see "A World without People," by Gerald I. Cohen (Sec. 21.3).

Caution: This advice to use the first person must definitely be qualified. Some publications and some organizations have a specific policy forbidding the use of I *or* we. *When you write for such a publication or such an organization, you must, of course, adhere to their rules* (at least, until you have gained a position of sufficient eminence and power to be able to change the rules). In the meantime, don't be *unnecessarily* impersonal.

15. Writing: Grammar

15.1 An Unpleasant Word

In any treatise on technical reporting, we come inevitably to a subject that we may as well frankly call *grammar*. Technical students (and others too) automatically recoil at the mention of this subject. Many of them take refuge in the popular notion that a direct approach and a lively style are more important than correct grammar.

Now we do not mean to belittle the importance of directness and liveliness. But the fact remains that, other factors being equal, grammatical writing is easier to read and easier to understand than ungrammatical writing. And it is easier not only for the reader who knows the rules himself, but also for the reader who is ignorant of them. For most of the rules of grammar are not simply arbitrary, capricious dictums put out by some old fuddy-duddy in order to trip you up in quizzes; they are, rather, codifications of logical relationships.

You may gather that we are trying to make grammarians out of technical men. This would be an impossible task even if it were a desirable one. But since the principles of grammar will help you communicate information clearly and efficiently, you ought to know at least the more important ones.

Unfortunately, the usual approach to grammar depends heavily on the rote memorization of a lot of formal rules and technical terms. You were probably subjected to this sort of thing in high school or in a freshman composition class. If you are like most students, you resisted this instruction and you have forgotten much of what you learned. Does one have to know what *addend, augend,* and *summand* are to be able to add two numbers together? No. Similarly, is it necessary to learn technical grammatical terms (beyond a few basic ones) to write grammatically? We believe not.

Therefore in this chapter (and the next one, on punctuation) we have relied as much as possible on common sense and logic and rules of thumb. We have tried to keep technical grammatical terms to a minimum and have restricted ourselves to the important rules that technical writers seem to overlook most often.

Thus this is a guide to correct usage, but it is by no means a complete handbook. When you are in doubt about any matter not covered here, look it up in a composition handbook. If you are interested in the fine points, investigate H. W. Fowler's *Dictionary of Modern English Usage* (Oxford).

If you can't seem to make a construction come out right even with the aid of a handbook, dodge the issue. Tear up what you have written and start over again with an entirely different construction and a fresh approach.

In the remainder of this chapter and in the next one, you will find a number of examples of writing that we have condemned as being *unclear.* You may argue that some of these examples are only momentarily unclear; or that the general sense or logic of the whole passage clarifies them; or that anybody could understand them with a little study. Certainly, permanent misunderstanding is worse than temporary misunderstanding. But even momentary misconstruction is a serious handicap to the reader. You owe it to your reader to make your meaning immediately clear with a minimum of study on his part.

15.2 Vague Pronoun

A common fault of technical writers is the use of pronouns whose meaning is vague or confusing. This is a serious fault, because it can result in misunderstanding. *Every pronoun must stand for a word (or an idea) that has already been expressed; and what it stands for must be immediately evident.*

Worst of all, of course, is the pronoun whose meaning cannot be deduced at all.

> These observations do not, however, explain why e_2 does not rise above E_b, and *this* will be important to us.

What will be important? The fact that e_2 fails to rise above E_b, or the fact that our observations do not explain the failure? The sentence contains no clue.

The pronoun whose meaning is momentarily in doubt is less serious, but it slows the reader down, particularly if he has to think things out in order to arrive at the meaning.

> The aluminum-nickel-iron alloys have external magnetic energy values five times those of the best quench-hardening alloys. *These* are carbon-free alloys whose magnetic hardness is controlled by. . . .

Are *these* the aluminum-nickel-iron alloys or the quench-hardening alloys? If the reader happens to be a metallurgist, he knows that *these* stands for the aluminum-nickel-iron alloys, because they are the ones that are carbon-free. But you should never require your reader to do this sort of puzzling even if he is capable of it. Again,

> As the temperature falls, a compressive stress is exerted by the bezel on the glass because of *its* greater temperature coefficient.

Anybody with technical training can figure out that it is the bezel that has the greater temperature coefficient; but why impose this task on your reader? A simple revision will get rid of the ambiguity and the passive construction at the same time:

> As the temperature falls, the bezel, because of its greater temperature coefficient, exerts a compressive stress on the glass.

Still less serious is the vague pronoun whose meaning is revealed as the reader gets farther on in the sentence. But even this sort of vagueness makes the reader's task more difficult, and you should always avoid it. The trouble is aggravated when the same pronoun is allowed to stand, in close succession, for two different antecedents. Here are some examples:

Original Example	*Suggested Revision*
Laminar flow has no corrosive power. *It* is seldom used because *it* is not economical to use low speeds.	Laminar flow has no corrosive power. It is seldom used because low speeds are not economical.
This intense specialization is necessary in our industrial society because of *its* complexity. *It* is advantageous because *it* allows an individual to become really proficient in at least one field.	This intense specialization is necessitated by the complexity of our industrial society. It is advantageous because it allows an individual to become really proficient in at least one field.
It became listless during the hot part of the day and revived when *it* got cooler.	It became listless during the hot part of the day and revived when the temperature fell.
This presupposes that the computer is working with continuously recycling problems during the check; but since no intelligible results can be obtained	This presupposes that the computer is working with continuously recycling problems during the check; but since no intelligible results can be obtained

Original Example	*Suggested Revision*
from the check unless *this* is being done, *this* is no limitation.	from the check unless the problems are recycling, this requirement is no limitation.
This is an unfortunate situation, and *one* exceedingly unpleasant for the honest scientist, and *one* is inclined to feel that the scientist should take no part in such proceedings.	This is an unfortunate situation, and one exceedingly unpleasant for the honest scientist. We believe that the scientist should take no part in such proceedings.
Attempts to build high-compression engines that would operate on low-octane gasoline depended on the theory that if maximum surface temperatures in the combustion chamber could be reduced by 600–1000 F, the mixture temperature could be controlled so that spontaneous ignition would not occur under high compression pressure. *This* required elimination of the exhaust valve, which reaches temperatures as high as 1300 F in automobile engines operating at maximum power.	Attempts to build high-compression engines that would operate on low-octane gasoline depended on the theory that if maximum surface temperatures in the combustion chamber could be reduced by 600–1000 F, the mixture temperature could be controlled so that spontaneous ignition would not occur under high compression pressure. *Such a reduction in surface temperature* required elimination of the exhaust valve, which reaches temperatures as high as 1300 F in automobile engines operating at maximum power.

The purist says that every pronoun must stand for a noun antecedent. In technical writing we can take a more liberal view and let pronouns stand for whole ideas that have been expressed if the reference is perfectly clear. In the following example, the pronoun *this* obviously stands for the whole idea expressed in the sentence that precedes it. To replace *this* by a wordier construction would be pedantic.

> Although the U.S. has a highway death toll much higher than that of any other country, our *death rate* per 100 million vehicle-miles is one of the lowest in the world. The reason for this is that the U.S. has by far the largest number of automobiles.

But remember: be sure your reader will be able to tell immediately and without pondering what every pronoun stands for.

15.3 Order of Sentence Elements

An important factor in clear writing is the proper order for the elements that go to make up each sentence. In general, try to *keep operators near the words they operate on.* If you do not, they will seem to operate on the wrong words. Here are some examples:

Original Example	*Comments*	*Suggested Revision*
A firm grip on the instrument and a firm support on the jaw of the patient are necessary when the mobilometer is being used to insure correct position, pressure, and steadiness during readings.	. . . the mobilometer is being used to insure correct position . . . ?	To insure correct position, pressure, and steadiness during readings, a firm grip on the instrument and a firm support on the jaw of the patient are necessary when the mobilometer is being used.
This rejection rate indicates that steps will have to be taken to reduce the incidence of short circuits by the manufacturer.	. . . short circuits by the manufacturer?	This rejection rate indicates that the manufacturer will have to take steps to reduce the incidence of short circuits.
Problems of a topographical nature are not likely to arise because the ground is generally flat.	. . . to arise because the ground is . . . flat?	Because the ground is generally flat, problems of a topographical nature are not likely to arise.
Purpose of the experiment was to compare half-wave potentials obtained experimentally for copper and lead in tartrate solutions with accepted literature values.	. . . solutions with accepted literature values?	Purpose of the experiment was to compare with accepted literature values the half-wave potentials obtained experimentally for copper and lead in tartrate solutions.
The use of large amounts of material and generous time allowances make macro analysis an excellent method of testing for persons with little training.	. . . testing for persons with little training?	The use of large amounts of material and generous time allowances make macro analysis a method of testing excellent for persons with little training.
The sample to be analyzed first must be put into solution.	. . . to be analyzed first?	The sample to be analyzed must first be put into solution.
Draw a line through the magnetization curve from the origin having this slope.	. . . the magnetization curve from the origin? . . . the origin having this slope?	Draw a line having this slope from the origin through the magnetization curve.
The effect of vibration and shock is difficult to report for permanent-magnet materials.	. . . to report for permanent-magnet materials?	The effect of vibration and shock on permanent-magnet materials is difficult to report

Original Example	*Comments*	*Suggested Revision*
The librarian will obtain books that are out of the library for the engineer.	. . . out of the library for the engineer?	The librarian will obtain for the engineer books that are out of the library.
In order to obtain reliable recordings on magnetic tape it appears that there are two factors that can cause loss of information that must be overcome.	. . . information that must be overcome?	Apparently there are two factors that can cause loss of information from magnetic tape. They must both be overcome in order to obtain reliable recordings.
No read-back pulses were obtained by either oscilloscope or ear phones on one channel using 300 feet of tape that could be attributed to tape imperfections.	. . . tape that could be attributed to tape imperfections?	No read-back pulses that could be attributed to tape imperfections were obtained by either oscilloscope or ear phones on 300 feet of one channel.
When the pulse arrives that de-energizes the rotating standard, the relay is opened.		Upon the arrival of the pulse that de-energizes the rotating standard, the relay is opened.

In particular, try not to separate a verb too far from its subject or its object or complement.

No trouble with engines running 1000 hours where valves and air cleaners were concerned has developed.	No trouble . . . has developed. What kind of trouble? Where valves and air cleaners. . . .	No valve or air-cleaner trouble has developed with engines running 1000 hours.

If the object or complement of a verb is compound, the apparent separation can be reduced by putting the shorter element nearer the verb:

The use of a-c propulsion with a synchronous motor requires a number of complicated interconnections between units and elaborate control equipment.	. . . interconnection between units and elaborate control equipment?	The use of a-c propulsion with a synchronous motor requires elaborate control equipment and a number of complicated interconnections between units.

Original Example	*Comments*	*Suggested Revision*
The new designs were at once suitable for production in the large quantities required and of improved performance.		The new designs were at once of improved performance and suitable for production in the large quantities required.

The word *only* is very frequently misplaced. In speech or informal usage it is customarily put before the verb. But in technical writing, precision requires it to go immediately before the word it modifies.

Original Example	*Comments*	*Suggested Revision*
This group will only study those problems that require lengthy investigations.	Seems to say that the group will do no more than *study* those problems . . . ; it will not do any actual experimental work.	This group will study only those problems that require lengthy investigations.
Chittagong only competes with Calcutta in the importing of mineral oils.	Could mean that Chittagong competes *only with* Calcutta . . . or that *only Chittagong* competes with Calcutta. . . . The actually intended meaning is made clear in the right-hand column.	Chittagong competes with Calcutta only in the importing of mineral oils.

15.4 Dangling Modifier

The dangling modifier does not often cause the downright confusion of the vague pronoun, but it is highly illogical, and on many readers it creates the same impression as the word *ain't*. Therefore you should avoid it.

Perhaps you have forgotten what a dangling modifier is. In general, it is a modifier that seems, at least momentarily, to operate on the wrong word. It is usually a verb form (often a participle) that is not supplied with a subject in its own sentence element, and that seems to claim a wrong word as its subject. An unfortunate order of sentence elements is often a contributing cause.

Consider this sentence, which does *not* contain a dangling modifier:

> By specifying standardized resistors, the designer can reduce the cost of the chassis.

The word *specifying* in the first part of the sentence is a verb form (it happens to be a gerund, but no matter) with no subject expressed. In the second part of the sentence *the designer* is the subject of *can reduce*. But it also seems to be the subject of *specifying;* and indeed it *is* logically the subject of *specifying*. Who does the specifying? The designer, of course. There is nothing wrong with this construction.

Now consider a similar sentence:

> By specifying standardized resistors, the cost of the chassis can be reduced.

Now *the cost* is in exactly the same situation as *the designer* was in the first sentence. As a result, *the cost* seems momentarily to be the subject of *specifying*. Does the cost do the specifying? No; some unnamed agent does. In this construction *specifying* is said to dangle. It is unattached, but it seems to attach itself to *the cost*.

You can usually correct a dangling modifier in several ways:

(1) By supplying a subject:

> By specifying standardized resistors, *we* can reduce the cost of the chassis.

(2) By changing the verb form to a word that is obviously a noun, and that therefore does not want to claim a subject:

> By *the specification of* standardized resistors, the cost of the chassis can be reduced.

(3) By changing the order of the sentence elements:

> The cost of the chassis can be reduced by specifying standardized resistors.

This construction leaves *specifying* still technically unattached, and some scholars may object to it. But because there is no noun in a position where it seems to be claimed as subject, the sentence is not even momentarily illogical or misleading.

(4) By rewriting the sentence completely:

> The use of standardized resistors will reduce the cost of the chassis.

The dangling modifier can assume various forms. The following examples should help you to recognize and avoid this blunder.

Original Example	*Comments*	*Suggested Revision*
During this period, besides writing and checking the program, four runs were made.	Did the four runs write the program?	During this period, the program was written and checked, and four runs were made.

Original Example	*Comments*	*Suggested Revision*
These resistors serve as a voltage divider of the proper size for observing the applied voltage step by coupling directly to the oscilloscope plates.	Do the resistors couple . . . ? Does the voltage divider observe?	Coupled directly to the oscilloscope deflector plates, these resistors serve as a voltage divider of the proper size for observation of the applied voltage step.
A bridged-T circuit was used. When very close to balance, the nonsinusoidal harmonics made the balance point indefinite.	Were the nonsinusoidal harmonics close to balance?	A bridged-T circuit was used. When it was very close to balance, the nonsinusoidal harmonics made the balance point indefinite.
Having made the decision, the space allotted for the installation must be large enough.	Did the space make the decision?	The decision having been made, the space allotted for the installation must be large enough.
The sentence gains in simplicity and clarity by eliminating superfluous words.	Does the sentence eliminate words?	The sentence gains in simplicity and clarity by the elimination of superfluous words.
This load curve is plotted so as to show the increasing slope with increasing load. When plotted in this way, additional curves may be drawn parallel to the original curve.	Are the additional curves plotted in this way? Only incidentally; primarily, it is *this load curve* that is plotted in this way.	This load curve is plotted so as to show the increasing slope with increasing load. When it is plotted in this way, additional curves may be drawn parallel to the original curve.
Of itself, the band theory of solids offers an incomplete explanation of ferromagnetism, but when considered jointly with the electron theory, many of the problems confronting us in our investigation can be answered.	Are the problems confronting us considered jointly with the electron theory? No; the band theory of solids is.	Of itself, the band theory of solids offers an incomplete explanation of ferromagnetism, but when *it is* considered jointly with the electron theory, many of the problems confronting us in our investigation can be answered.
The curves for an amplifier will indicate the range of voltage over which linear assumptions are justified just by looking at them.	Do the curves look at them (selves)?	A glance at the curves for an amplifier will indicate the range of voltage over which linear assumptions are justified.

Original Example	*Comments*	*Suggested Revision*
The meter is not sufficiently sensitive to be able to read the small differences involved.	Should the meter be able to read the differences?	The meter is not sufficiently sensitive *for us* to be able to read the small differences involved. *or:* The meter is not sufficiently sensitive *to show* the small differences involved.
In the neighborhood of the crystal's resonant frequency, P. Vigoureaux has derived the following values for the parameters.	Was P. Vigoureaux in the neighborhood of the crystal's resonant frequency?	P. Vigoureaux has derived the following values for the parameters in the neighborhood of the crystal's resonant frequency.
A high-voltage supply must be used to be able to change the coil currents rapidly.	Is the high-voltage supply able to change the currents?	A high-voltage supply must be used to permit the coil currents to be changed rapidly.
After locating the site, it did not fulfill expectations.	Did it (the site) locate the site?	After being located, the site did not fulfill expectations.
After calibrating the manometer, three patients were found to have high blood pressure.	Did the patients calibrate the manometer?	After the manometer had been calibrated, three patients were found to have high blood pressure.
With the baffle in place, fuel does not splash when flying in rough air.	Is the fuel flying in rough air?	. . . fuel does not splash when the airplane is flying in rough air. *or:* . . . fuel does not splash during flight in rough air.
Combining the two adjustments properly, the jaw speed can be varied from 0.002 inch per minute . . .	Does the jaw speed combine the adjustments?	With the proper combination of the two adjustments, the jaw speed can be varied from 0.002 inch per minute . . .

15.5 Split Infinitive

You have probably been taught that the split infinitive is a particularly bad fault. You should try to avoid this construction, but not at the expense of (1) reducing clarity or (2) sounding forced and artificial.

Here is a perfectly clear sentence that contains a split infinitive.

> A study must be included to properly integrate the computer with the other main components of the control system.

You might move the splitting word—*properly*—to any of several different places:

> (1) A study must be included properly to integrate the computer . . .

Now *properly* might modify *included*. This revision must be rejected because it make the sentence ambiguous.

> (2) A study must be included to integrate properly the computer with the other main components of the control system.

This sounds strange and foreign because it simply is not English idiom. It must be rejected.

> (3) A study must be included to integrate the computer with the other main components of the control system properly.

Now *properly* is so far removed from *integrate* that it seems to hang by itself, unattached.

> (4) A study must be included to integrate the computer properly with the other main components of the control system.

This is the best of the four possible revisions, but it still lacks the clarity and the vigor of the original with the split infinitive.

Here is another clear sentence that contains a split infinitive:

> Consumption is expected to more than double by 1985.

The words in this sentence cannot go in any other order and still say the same thing. To avoid this split infinitive—if you must—you will have to rewrite the sentence entirely.

Nowadays most authorities take a moderate view of the split infinitive. You should not hesitate to use it when it is the clearest, most natural way of expressing your meaning.

But you can often avoid the split infinitive with no difficulty. Consider this sentence:

> This course should help the engineer to *more quickly and effectively* handle a number of situations than he can if left to his own devices.

The splitting element can be moved either toward the beginning of the sentence or toward the end. Generally, if you can move the splitting element toward the *end* of the sentence without beclouding your meaning, the revision will be a successful one. But if, as in the original example below, you move the splitting element toward the beginning of the sentence, the result is likely to sound artificial and perhaps ambiguous.

Original Example	*Suggested Revision*
This course should help the engineer *more quickly and effectively* to handle a number of situations than he can if left to his own devices.	This course should help the engineer to handle a number of situations *more quickly and effectively* than he can if left to his own devices.

Here are several split infinitives susceptible of easy repair:

It is possible to so construct such an amplifier that its behavior is almost independent of frequency over a wide range.	It is possible to construct such an amplifier so that its behavior is almost independent of frequency over a wide range.

But notice the artificial, stilted sound of

It is possible so to construct such an amplifier . . .

An adjustment is provided to intermittently check the calibration of the gauge.	An adjustment is provided to check the calibration of the gauge intermittently.

Notice the ambiguity of

> An adjustment is provided intermittently to check the calibration of the gauge.

Original Example	*Suggested Revision*
. . . to more easily visualize just what happens to the money that is invested.	. . . to visualize more easily just what happens to the money that is invested.
. . . to sketchily mention all of the minute procedures . . .	. . . to mention sketchily all of the minute procedures . . .

Finally, remember that the split infinitive is specifically the separation of a verb from the auxiliary *to*. Verbs appear also with other auxiliaries, from which you may separate them with no qualms at all, so long as you do not separate them too far. For instance, *to always come* is a split infinitive; *have always come, do always come, will always come* are not. Similarly, *to forever be going* is a split infinitive; *to be forever going* is not.

15.6 Tense

You have probably been warned never to switch tenses in the middle of the stream. In general this is good advice; certainly you should not switch tenses needlessly.

But logic and meaning sometimes require two different tenses in the same sentence. This situation often arises in technical writing in the description of an event that took place in the past but that depended on some principle or fact or relationship that continues to hold in the present. Thus we may better modify the rule and say: *Be sure that every change of tense carries some significance.*

For instance, in order to maintain consistency of tenses you might write:

> The balloon rose because helium was less dense than air.

This sentence implies that helium used to be less dense than air, but that it isn't any more. That is manifestly absurd. To make your meaning clear, you must mix tenses within the sentence:

> The balloon rose because helium is less dense than air.

But complications arise when tenses are shifted meaninglessly. Here are some examples:

Original Example	*Suggested Revision*
Tests *have shown* that the heat contents of various gasolines *differ* by about 3 percent in Btu per pound. But when the same fuels *were* evaluated in terms of Btu per gallon, the differences *were* as high as 15 percent.	Tests *have shown* that the heat contents of various gasolines *differ* by about 3 percent in Btu per pound. But when the same fuels *are* evaluated in terms of Btu per gallon, the differences *are* as high as 15 percent.

The revised version still contains two different tenses; the switch does carry some meaning.

The first charge breaks part of the material and forms a weak point. The following charges expend more of their energy in the direction of this weak point than they would have if they had been fired at the same time as the first charge.	The first charge breaks part of the material and forms a weak point. The following charges expend more of their energy in the direction of this weak point than they would if they were fired at the same time as the first charge.

15.7 Parallelism

Logic and orderliness demand parallel grammatical constructions for parallel ideas. More important, parallel constructions will make your meaning clearer because a changed (or nonparallel) construction hints to the reader that the subject has been changed too.

Original Example	*Suggested Revision*
In compression the pieces broke off in a plane 45 degrees to the long axis. The tensile fracture was perpendicular to the long axis.	In compression the pieces fractured in a plane 45 degrees to the long axis. In tension they fractured perpendicular to the long axis.

In the original version the difference between the fractures is obscured by the difference between the sentences. In the next two examples, faulty parallelism leads to real momentary confusion because the parallel ideas seem to merge rather than to stand separately.

If the meter under test is fast, the dial reading will be less than 1.0; and more than 1.0 if it is slow.	If the meter under test is fast, the dial reading will be less than 1.0; if it is slow, the dial reading will be more than 1.0.
Titles of articles or chapters in printed matter should be in italics and underlined when typed.	Titles of articles or chapters in printed matter should be in italics; in typed matter they should be underlined.
Executives in the Sales Division should be *marketing-conscious* [adjective] rather than *good production men* [noun].	Executives in the Sales Division should be *salesmen* rather than *production men*.
The customer would *ask* for a certain brand rather than *asking* for recommendations.	The customer would *ask* for a certain brand rather than *ask* for recommendations.
The testing of these meters involves a comparison of the speed of the meter disc under test with the disc speed of a secondary meter standard.	The testing of these meters involves a comparison of the disc speed of the meter under test with the disc speed of a secondary meter standard.
He will also learn the correct pronunciation and that there are two ways of spelling the word.	He will also learn the correct pronunciation, and he will find that there are two ways of spelling the word.

15.7.1 Lists

Faulty parallelism is particularly obvious and particularly annoying in any kind of list or enumeration.

Original Example	*Suggested Revision*
The shielding for the system is provided by the *use* of coaxial cable to connect sensing elements, *insertion* of r-f chokes in the power supply, and *enclosing* the tank circuits in shielded cans.	The shielding for the system is provided by the *use* of coaxial cable to connect sensing elements, *insertion* of r-f chokes in the power supply, and *enclosure* of the tank circuits in shielded cans.
1. Nonequilibrium conditions. 2. Sampling. 3. Excessive heat. 4. Other elements present may have interfered.	1. Nonequilibrium conditions. 2. Sampling. 3. Excessive heat. 4. Possible interference by other elements present.
A knowledge of the carburetor air temperature is needed *for adjustments* of throttle position and *to warn* of icing conditions.	A knowledge of the carburetor air temperature is needed *for adjustments* of throttle position and *for warning* of icing conditions.
Smith pointed out *how* Jones opposed every innovation, *how* he encouraged the men to slow down, *that* he never took the initiative, and *that* he could not keep the machines running.	Smith pointed out *that* Jones opposed every innovation, *that* he encouraged the men to slow down, *that* he never took the initiative, and *that* he could not keep the machines running.
The assumptions made are: (1) the accuracy of the measuring means is greater than the accuracy of the instrument being measured; (2) the noise level and drift are less than the inaccuracy of linearity; and (3) *that* the points of maximum deviation from the straight line are known.	Omit *that*.

15.7.2 Bastard Enumeration

You are familiar with the series form, or enumeration, *a, b, and c,* in which the letters stand for elements of almost any grammatical form—words, phrases, clauses, or even sentences. (For the punctuation of the series see Sec. 16.2.4.) As shown in Sec. 15.7.1, all of the elements in any one series must be of the same grammatical form.

Further, remember that *any external operator that operates on more than one element of a series operates automatically on* all *elements of the series.* Disregard of this rule produces a special form of nonparallel construction that Fowler calls the *bastard enumeration* (*Modern English Usage*). The bastard enumeration looks like a legitimate enumeration but isn't one. It is a subtle fault that plagues many good writers. Because it is an offense against logic and clarity, you should avoid it.

Consider this sentence:

> Adapters are made in various sizes, shapes, weights, and with any number of leads.

The elements of this enumeration are (*a*) sizes, (*b*) shapes, (*c*) weights, and (*d*) with any number of leads. The words *in various* obviously operate on *a, b,* and *c;* they therefore operate also on *d*. The result is . . . *in various with any number of leads*.

The cure for this error is often the insertion of another *and,* which separates the bastard enumeration into two parts, one of which is a true enumeration:

> Adapters are made in various sizes, shapes, *and* weights, and with any number of leads.

Now *with any number of leads* is not an element of the series; and *in various* operates only on *sizes, shapes,* and *weights*.

This bastard enumeration may be corrected in another way:

> Adapters are made in various sizes, in various shapes, in various weights, and with any number of leads.

Now *in various* is not an external operator; it is repeated within each of the first three elements of the enumeration and so does not have any connection with the fourth.

Here are some other examples of the bastard enumeration:

Original Example	*Comments*	*Suggested Revision*
When in doubt about presentation, style, expression, or tempted by figures of speech, rush to the editor.	Since *tempted by figures of speech* seems to be a member of the series, *when in doubt about* operates on it, too.	When in doubt about presentation, style, or expression, or when tempted by figures of speech, rush to the editor.
When maintained at the appropriate pH, osmotic pressure, and in the presence of certain salts . . .	*Maintained at* operates on *appropriate pH* and *osmotic pressure;* therefore also on *in the presence of certain salts.*	When maintained at the appropriate pH *and* osmotic pressure, and in the presence of certain salts . . .
The tape equipment is undergoing test to determine operating characteristics, relay timing, power requirements, fusing, and to suppress undesirable transients.	Since *to suppress undesirable transients* seems to be a member of the series, *to determine* operates on it.	The tape equipment is undergoing tests to determine operating characteristics, relay timing, power requirements, *and* fusing, *as well as* to suppress undesirable transients.
March 29, 30, 31, and April 1.		March 29, 30, and 31 and April 1.

Original Example	*Comments*	*Suggested Revision*
The thiomorpholide was extracted with ether, washed with 10-percent sodium carbonate, 10-percent hydrochloric acid, and with water.	1. *Extracted with ether* is fused into the series that starts with *washed with.* 2. *Washed with* operates on *10-percent sodium carbonate and 10-percent hydrochloric acid,* and therefore also on *with water.*	The thiomorpholide was extracted with ether *and* washed with 10-percent sodium carbonate, 10-percent hydrochloric acid, and water.

15.8 Preposition at End of Sentence

At some time during your schooling you may have been taught that no sentence should end with a preposition. This view is now held only by pedants. Good present-day writers end sentences with prepositions whenever that is the natural word order. The sentence that has been revised to avoid a preposition at the end is often stilted or awkward.

Rejoice: here is one "rule" that you can just forget about.

15.9 Number with Collectives

A noun that denotes a whole made up of a number of similar parts is called a collective. Common examples are: pair, set, group, majority, 30 inches, two million dollars. Notice that some of these words are singular in form, others plural.

A collective may take either a singular verb or a plural verb, depending upon its sense. When it refers to the whole group as a unit, the collective takes a singular verb. When it refers to the separate entities that go to make up the group, the collective takes a plural verb. The number of the pronoun that stands for a collective must, of course, agree with the number of the verb.

This rule is commonly overlooked. Many writers stick slavishly to singular verbs with all collectives that are singular in form and to plural verbs with all collectives that are plural in form. You can help your reader by making the number of the verb fit the sense of the collective rather than its form.

15.9.1 Apparently Plural Words

Would you say "Two million dollars are a lot of money"? No, you would say "Two million dollars *is* a lot of money." You are referring not to the separate dollars, but to the whole sum of money. Yet technical writers often becloud their meaning by using plural verbs in this construction:

Original Example	*Suggested Revision*
Ten grams of the isotope *were* collected.	Ten grams of the isotope *was* collected.
To 200 grams of sodium *were* added 1.5 kilograms of crushed ice.	To 200 grams of sodium *was* added 1.5 kilograms of crushed ice.
In the forward hold there *are* 16 inches of water. *It* [!] must be pumped out.	In the forward hold there *is* 16 inches of water. *It* must be pumped out.

The writers of these examples were referring not to separate grams and kilograms and inches, but to a quantity of isotope and a quantity of ice and a quantity of water measured in grams and kilograms and inches. The singular verbs make sense; the plural verbs do not.

On the other hand you would, of course, say "A million pennies *were* scattered around the vault."

15.9.2 Apparently Singular Words

You have probably heard an exclamation like "A pair of jacks *is* the highest hand I've seen for an hour." The unlucky gamester is referring not to the separate jacks, but to the pair as a unit. The singular verb makes sense. But now consider this statement:

> A pair of hawks was taking turns feeding the young birds.

How can a single entity (as implied by the singular verb) take turns? A clearer and more logical statement is:

> A pair of hawks were taking turns feeding the young birds.

The singular verb with a collective that is plural in sense though singular in form can produce a construction that is both illogical and inconsistent, as in the following example.

Original Example	*Suggested Revision*
In beta-decay some of the neutrons already in the nucleus are changed to protons, and an equal number of electrons *is* emitted by the nucleus.	In beta-decay some of the neutrons already in the nucleus are changed to protons, and an equal number of electrons *are* emitted by the nucleus.

Here are some other examples of plural verbs (and pronouns) used logically with collectives that are plural in sense though singular in form:

> The *majority were* between 1.5 and 2.5 millimeters long; *they* moved constantly and randomly. The *rest were* mostly smaller; *they were* also less active.

A number of ill-concealed snickers *were* heard in the back of the lecture hall.

15.9.3 Data

The word *data* used to be defined only as *plural of datum*. It meant *facts* or *figures*. To use it with a singular verb was considered highly improper.

Over a comparatively short span of years *data* has come to mean also *information*. In this sense it is logically singular. Thus *data* is a collective, and the number of its verb may be either singular or plural, depending upon whether it means *a body of information* or a lot of *separate figures*.

Some authorities still insist that *data* must never appear with a singular verb (or pronoun). But unless your boss objects, you may use *data* as singular or plural according to its meaning.

In the first example below, the singular would not make sense. In the second, the plural would sound very stilted.

> The data *were* plotted point by point. Some of *them were* found to be grossly in error.
>
> The data developed by the Senate investigators *is* going to be published in full next year. *It* should be helpful to anyone working in the field of criminology.

15.10 Nouns Used as Adjectives

The *attributive noun*—a noun used as an adjective—is an entirely legitimate construction. However, in a worthy effort to eliminate a few *of*'s and *the*'s, many engineers use attributive nouns much too freely. The trouble is that this construction can be confusing, at least temporarily, especially when a number of attributive nouns are strung together into a compound adjective preceding the noun they modify.

When several *adjectives* (perhaps with adverbs) precede a noun, the reader is aware from the beginning that they will ultimately modify a noun that is to come later, as in

> a highly exorbitant, entirely unacceptable cost

But now consider this example, with its compound modifier made up of attributive nouns:

> It is necessary to eliminate part of the triple superphosphate plant waste disposal cost.

The function of the phrase *triple superphosphate plant waste disposal* is not evident until we reach the word it modifies, *cost*. The relationship is quite vague. Much clearer is:

> It is necessary to eliminate part of the cost of waste disposal at the triple superphosphate plant.

Notice that in their new positions, *disposal* and *plant* no longer act as adjectives, but as ordinary nouns.

Here is an expression that actually appeared in a treatise on control systems:

> reference quantity roughness amplitude-dependent quantity roughness amplitude ratio.

After considerable study, you will discover that *ratio* is modified by a long compound adjective composed of all the words that come before it. As stressed in Sec. 16.7.2, compound adjectives that precede the nouns they modify should be hyphenated. And hyphenation does help to clarify an expression like that one:

> reference-quantity-roughness-amplitude / dependent-quantity-roughness-amplitude ratio.

But even when hyphenated, such a compound is very confusing because the relationships of the earlier words do not become apparent until you get to the later words. The same idea can be stated quite clearly if the modifier follows the noun:

> ratio of reference-quantity roughness amplitude to dependent-quantity roughness amplitude.

Here are a number of examples of temporarily confusing attributive-noun phrases that are clarified by being moved to a position later in the sentence, with the addition of a few *of*'s and *the*'s:

Original Example	*Suggested Revision*
The fuel injection systems used for aircraft were controlled by engine speed-air density responsive metering systems.	The fuel injection systems used for aircraft were controlled by metering systems responsive to engine speed and air density.
Motor vehicle standard cost estimating system study	Study of systems for estimating cost of motor vehicle standards
Transportation system safety assurance demands a systematic effort.	Only a systematic effort will assure safety in our transportation system.
Special attention will be given to promotion of upper torso restraint system use.	The use of upper torso restraint systems will be specially promoted.

Original Example	*Suggested Revision*
Drug efficacy study implementation	Implementation of the study of drug efficacy
Rear lighting perception ability	Ability to perceive rear lighting
Steering system component degradation	Degradation of steering system components

The compound does not have to be as long as those to be improved by a shift to the rear.

Original Example	*Suggested Revision*
1½-inch-wide lining	lining 1½ inches wide
air-carbon dioxide mixture	mixture of air and carbon dioxide
air-fuel gas mixture	mixture of air and fuel gas
tire-failure prediction	prediction of tire failure
simplified circuit construction (ambiguous)	construction of simplified circuits

15.11 Spelling

There is little inherent virtue in spelling correctly. But spelling errors, like some other apparently trivial errors, have much the same effect as saying *ain't*: they make the reader wonder whether the writer is educated; whether, indeed, he knows what he is talking about. You owe it to yourself to spell correctly.

Whenever you have a serious writing job to do, be sure to keep a dictionary close by and consult it freely. (For a recommended dictionary see Sec. 14.2.5.)

If you find that you are inclined to make spelling errors, keep a list of the words you have spelled wrong and look at it from time to time. If you are constitutionally unable to learn spelling (we have met a few people who claim to be), the least you can do is to have some capable friend go over everything you write before you submit it to the people who count.

Here are some principles of spelling that most people find to be of considerable help.

(1) A final consonant is doubled before an ending that begins with a vowel when the word to be obtained is accented on the final syllable of the root word.

prefer	preferred
prefer	preference
occur	occurrence

(2) When words contain *ei* or *ie*, *i* usually comes before the *e*, except when the combination comes after *c*.

believe	retrieve	receive

(3) Some poor spelling results from careless pronunciation. If, in conversation, you leave off syllables or add them where they do not belong, you may spell these words as given in the *wrong* column.

Wrong	*Right*
accidently	accidentally
athelete	athlete
preform	perform
perserverance	perseverance

(4) We should add a note here on *preferred* spelling. Some British spellings of words have gone through a process of *Americanization*. In most technical writing, the American version is preferred. When you know that two versions exist but you don't know which is preferred, consult your dictionary.

Here are some examples of words that have changed their spelling:

American	*British*
acknowledgment	acknowledgement
judgment	judgement
maneuver	manoeuvre
practice (verb)	practise
theater	theatre

(5) Three pairs of words whose members are pronounced alike but spelled differently cause a lot of spelling errors in technical writing. These words are distinguished below (only the meanings that are commonly confused are given).

(a) *Principle:* Noun = basis, fundamental truth, basic law
Principal: Adjective = chief, foremost

(b) *Effect:* Noun = result produced; Verb = bring about, accomplish
Affect: Verb = influence; have an *effect* on

(c) *Foreword:* Noun = preface (front *word*)
Forward: Adjective, adverb = toward the front (front *direction*)

The technical writer should be on his guard about some words that will appear frequently in his writing:

accommodation	judgment	personnel
acknowledgment	laboratory	precede
analogous	maintenance	proceed
carburetor	occurred	procedure
cylinder	occurrence	propeller
exceed	questionnaire	schedule
guarantee	pamphlet	superintendent

16. Writing: Punctuation

16.1 A Help to Your Reader

Proper punctuation performs an indispensable function: it helps to make the writer's meaning precise and clear; and it eases the reader's task. Conversely, slipshod punctuation can actually alter meaning; at the very least it puts the reader (perhaps many readers) to a lot of unnecessary trouble.

Possibly because of the insignificant physical aspect of punctuation marks, some people consider the whole subject trivial and beneath their notice. Others are careless about punctuation on the plea that they like the "open style" or the smooth-flowing effect associated with a general absence of punctuation.

This attitude may be appropriate for some kinds of writing; but remember that the primary aim of technical reporting is to transmit information accurately and efficiently. Since punctuation can play an important part in this process, you should take the trouble to punctuate your technical writing fairly rigorously. If you don't like the looks or the effect of a lot of commas, the way to get rid of them is not to omit clarifying punctuation, but to construct your sentences so that little punctuation is needed.

Most technical students have had elementary training in composition that was directed to all the students in a high school class or a college freshman class without regard to their future specialization. Furthermore, the teachers of these classes have probably had a literary rather than a technical background. Consequently some of the principles propounded in *Technical Reporting* may differ from those you have been taught before. Since this book is designed for the technical writer, you will probably do well to heed its advice whenever your primary aim is precision and clarity rather than literary effect.

You can learn to do a reasonably good job of punctuating without memorizing a large number of formal rules. The remainder of this chapter presents some general principles and some rules of thumb that should help you. It stresses only a few highly important formal rules that are commonly overlooked. If you need further information on punctuation, consult a composition handbook or look up the excellent articles on punctuation in one of the dictionaries listed in Sec. 21.3.

16.2 The Comma

16.2.1 General Function of the Comma

The comma is a separator, or pause-indicator. It separates words or parts of sentences that would otherwise run together. It is the lightest member of the series that consists of the period, the semicolon, and the comma. It tells the reader those things that a speaker indicates to his listeners by pausing or dropping his voice.

This suggests a rule of thumb for the use of the comma. Whenever you wonder whether to insert a comma, read your sentence aloud, with exaggerated expression. Generally speaking, where the tone of your voice changes or where you pause momentarily, a comma belongs.

Read these two sentences aloud, slowly:

> It is therefore important for every engineer to be a man of honor.
> It is, therefore, important for every engineer to be a man of honor.

Notice that you go straight through the first sentence without pause. But in the second sentence you pause before and after *therefore;* you say *therefore* in a slightly subdued tone; and you give a little stress to *is*.

There is no question here of "correct" punctuation; both sentences are correctly punctuated. The commas in the second sentence simply change the emphasis. Often, of course, commas play a much more vital role: by indicating separations, they actually control meaning.

You can usually decide whether to use a comma on the basis of this simple voice-drop test. But the functions of the comma described in Secs. 16.2.2–16.2.8 are so important (and so often overlooked) that you should learn to recognize them and the rules that govern them. *The rules in Secs. 16.2.2, 16.2.3, and 16.2.4 are particularly important; you should probably follow them formally and rigorously.* Notice, however, that the voice-drop criterion fits in with these rules, too.

16.2.2 The Comma in Compound Sentences

The compound sentence is usually a sentence made up of two independent clauses (i.e., independent sentences) joined by a coordinating conjunction such as *and, or, but,* or *for.* Most grammar books say to put a comma before the conjunction unless the clauses are very short. This advice seems to be taken by most people to mean that the comma can usually be omitted.

If you want to be sure that you will never confuse your reader, even temporarily, you must practically always use a comma in this construction. The following examples illustrate the momentary misunderstanding that sometimes occurs when the comma is omitted.

Orginal Example	*Confusing Passage*	*Suggested Revision*
Throughout the islands move numerous small craft and larger vessels make regular calls at the larger ports.	. . . move numerous small craft and larger vessels . . .	Throughout the islands move numerous small craft, and larger vessels make regular calls at the larger ports.
A few modifications were made on the operation matrix drivers and the control-pulse output units will be modified.	. . . modifications were made on the operation matrix drivers and the control-pulse output units . . .	A few modifications were made on the operation matrix drivers, and the control-pulse output units will be modified.
Your schedules must be revised or corrected versions will not be ready for the next report.	. . . schedules must be revised or corrected . . .	Your schedules must be revised, or corrected versions will not be ready for the next report.
When a code is found, the control causes it to be set up in the relay register associated with the reader and the contents of the two relay registers are then checked for coincidence.	. . . associated with the reader and the contents of the two relay registers . . .	When a code is found, the control causes it to be set up in the relay register associated with the reader, and the contents of the two relay registers are then checked for coincidence.
Air brakes can give very rapid decelerations and the ability to reduce speed abruptly could be helpful in an approach system.	. . . give very rapid decelerations and the ability to reduce speed abruptly . . .	Air brakes can give very rapid decelerations, and the ability to reduce speed abruptly could be helpful in an approach system.

The insertion of a comma before the conjunction clarifies every one of these sentences. But most of them could be further improved by more extensive revisions, particularly by being changed from compound sentences to complex sentences, with one of the clauses subordinated (see Sec. 14.12). For instance, the third example would be more meaningful if it read:

> Unless your schedules are revised, correct versions will not be ready for the next report.

16.2.3 The Comma around Nonrestrictive Modifiers

The restrictive, or defining, modifier pins down its object to a certain one (or certain ones) from a larger class:

> Men seldom make passes at girls *who wear glasses.*
> —Dorothy Parker

We are not talking about girls in general; just that restricted group *who wear glasses*. The restrictive modifier is essential to the meaning of the statement; *it is never set off by commas.*

The nonrestrictive, or commenting, modifier, on the other hand, simply makes a comment about its object without limiting our choice:

> Grammar, *which is a dull subject,* is important.

Are we limiting ourselves to that part of grammar which is a dull subject? No; all grammar is dull; we are simply adding this comment. The nonrestrictive modifier is not essential to the meaning of a statement; *it should always be set off in commas.*

If you have difficulty in deciding whether a modifier is restrictive or nonrestrictive, try crossing it out. If what remains is still a sensible statement, the modifier is nonessential, or nonrestrictive. For instance,

> Grammar . . . is important.

is a perfectly sensible statement even if you don't agree with it. The modifier *which is a dull subject* should therefore be set off in commas. On the other hand, if what remains is no longer a sensible statement, the modifer is essential, or restrictive. For instance,

> Men seldom make passes at girls. . . .

is manifestly absurd. The modifier *who wear glasses* should not be set off by a comma.

The modifiers in the two examples above are both relative clauses, because it is the punctuation of relative-clause modifiers that seems to cause most of the difficulty. But the same rules apply to other kinds of modifiers, as in the following examples:

Reports *written by engineers for engineers* may contain algebraic expressions. (Restrictive.)

Reports, *written or oral,* should be as clear and as short as possible. (Nonrestrictive.)

Technical writers occasionally make the mistake of putting commas around restrictive modifiers, particularly long ones. Much more often, though, they omit the commas from nonrestrictive modifiers, a usage for which they can find ample precedent in the newspapers.

Some writers take the attitude that they will put commas around nonrestrictive modifiers whenever confusion would result without them. But unless you consistently put commas around *every* nonrestrictive modifier, you will sometimes leave your readers in doubt.

Suppose you are *not* rigorous about this rule. In a report you write:

> This drawing shows the end sections *which differ from the middle sections.*

Your reader wonders whether you purposely omitted a comma after *end sections.* If you did, *which differ from the middle sections* is restrictive, and you are talking about just those end sections which differ from the middle sections. But perhaps all the end sections are alike, and you are simply commenting that they differ from the middle sections. You thought a comma after *end sections* unnecessary. How is your reader to know which you intended? As a matter of fact, the writer of that sentence simply neglected to put a comma after *end sections,* where it was required by his meaning.

Here are two more examples that are ambiguous from a writer who is not rigorous about punctuating nonrestrictive modifiers:

> The filter acts as a diffusing screen for the air *which enters the top of the chamber.*
>
> This cell energizes the thyratron tube *which draws current through the coils.*

Again, the writers of these sentences were slipshod, and each of the relative clauses should have been preceded by a comma. You should get into the habit of putting commas around every nonrestrictive modifier, no matter how simple and obvious it may be.

16.2.4 The Comma in Series

Teaching differs on the punctuation of the series, or enumeration, of the form *a, b, and c.* Some schools say that the comma preceding the conjunction should be omitted; others that it should be omitted except where confusion would result; and still others that it should always be used. *Unless you put a comma before the conjunction in every series, you will sometimes confuse your readers.*

Suppose you write:

> Electric lenses were developed by Davisson and Calbrick, Knoll and Ruska and Bruche.

This is evidently meant to be a three-member series, *a, b,* and *c.* The first member, *a,* is obviously Davisson and Calbrick; but does Ruska belong with Knoll or with Bruche? As the sentence now stands, there is no clue.

Suppose you belong to the school that says the comma should go into this construction when confusion would result without it. Then you would write:

> . . . Davisson and Calbrick, Knoll and Ruska, and Bruche.

Remember, your reader has observed that you often omit the last comma in a series. When he comes to *Knoll and Ruska,* he thinks *momentarily* that the three members of the series are (*a*) Davisson and Calbrick, (*b*) Knoll, and (*c*) Ruska. When he gets to *and Bruche,* he sees that Ruska goes with Knoll as the *b* member of the series, and he understands you completely.

On the other hand, if he has observed that you never omit that last comma, he knows immediately that Ruska is not the last member of a series, and he is not misled even momentarily.

Now consider this passage:

> If we have subdivisions of *X, Y* and *Z,* whose values are known. . . .

Do we have three subdivisions—(1) of *X,* (2) of *Y,* and (3) of *Z?* Or do we have two subdivisions of *X* that we are calling respectively *Y* and *Z?* If this passage comes from a writer who consistently puts a comma before the conjunction in a series, we know that the three letters are not members of a series, and that *Y* and *Z* are subdivisions of *X*. But from a writer who is not consistent, how are we to tell? As a matter of fact, the man who wrote that passage intended to speak of subdivisions of *X, Y,* and *Z.*

Here is a similar sentence:

> A sampling device is made up of two cascaded components, a sampling switch and a cascaded holding device.

From a nonrigorous punctuator, this sentence might be momentarily interpreted in either of two ways: (1) a sampling device is made up of three elements—(*a*) two cascaded components, (*b*) a sampling switch, and (*c*) a cascaded holding device; or, (2) a sampling device is made up of two cascaded components—namely, a sampling switch and a cascaded holding device. On the other hand, if this passage had come from a rigorous punctuator, we would know at once that no series was intended, and that the words *a sampling switch and a cascaded holding device* are simply further explanation of *two cascaded components.* (To remove all doubt, use a dash instead of a comma after *components;* see Sec. 16.5.)

To be on the safe side, get in the habit of putting a comma before the conjunction of *every* series.

16.2.5 The Comma with Adjectives in Series

Two or more adjectives preceding the noun they modify constitute a special kind of series. If these adjectives modify the noun separately and independently, they should be separated from each other by commas. On the other hand, if the earlier adjectives modify not only the noun but also the succeeding adjectives, they should not be separated by commas. This rule can be expressed algebraically. Let *b, c, d* be adjectives; *x,* a noun.

$(b+c+d)\ x$ —Commas to separate *b, c, d.*

$b\ \left\{c\ [d(x)]\right\}$ —No commas.

For example you would write

a big, ungainly freshman.

Big and *ungainly* operate separately on *freshman.*

On the other hand, you would write

a big black dog.

Big modifies the whole concept of *black dog.*

Here is a rule of thumb to determine when adjectives in series should be separated by commas: Try putting the word *and* at each place where you think a comma might belong. If the passage still sounds sensible, use the comma; if it doesn't, omit the comma. If you are in doubt, omit the comma.

Application of this test to the examples above produces

a big and ungainly freshman,

which is sensible, and

a big and black dog,

which is silly.

Here are several other examples of adjectives in series. You can confirm the usage of commas by trying the *and* test.

Commas Indicated	*Commas Not Indicated*
A big, powerful, handsome motor	A wound-rotor alternating-current repulsion-induction motor
A badly worn, obsolete engine	A 4-cycle gasoline engine
A hasty, careless, inaccurate statement	A hasty rearward flight

For a discussion of the *compound adjective,* which is a combination of words rather than a succession of separate words, see Sec. 16.7.2.

16.2.6 The Comma after Introductory Elements

Section 16.2.1 points out that the comma is used to separate words or parts of sentences that would otherwise run together. The introductory phrase or clause (an adverbial phrase or relative clause that comes before the main clause) often seems to run on into the rest of the sentence, at least momentarily. You can avoid this source of confusion by putting a comma after every introductory phrase or clause (unless it is very short and the break between it and the main clause is obvious).

Original Example	*Confusing Passage*	*Suggested Revision*
When the potential is kept constant and the frequency increases the impedance of the voltmeter increases.	. . . frequency increases the impedance . . .	When the potential is kept constant and the frequency increases, the impedance of the voltmeter increases.
At the instant of starting the motor draws more than 400 amperes.	. . . starting the motor . . .	At the instant of starting, the motor draws more than 400 amperes.
When propagation time and switching time are allowed for the hypothetical system would probably require a pulse-repetition frequency of 20 kilocycles.	. . . allowed for the hypothetical system . . .	When propagation time and switching time are allowed for, the hypothetical system would probably require a pulse-repetition frequency of 20 kilocycles.
When the bulb is exposed to temperatures different from that of the medium in which the stem is placed a severe gradient exists across the bulb.		When the bulb is exposed to temperatures different from that of the medium in which the stem is placed, a severe gradient exists across the bulb.

Original Example	*Confusing Passage*	*Suggested Revision*
Whenever the computer is operating the control element is generating a succession of pulses.	. . . computer is operating the control element . . .	Whenever the computer is operating, the control element is generating a succession of pulses.

Several of these examples would be further improved by more extensive revision. The commas at least make them clear.

16.2.7 Second Comma of a Parenthetical Pair

The single comma is the lightest in the series of pause-indicators that consists of the period, the semicolon, and the comma. Similarly, the *pair* of commas is the lightest in the series of parentheses that consists of the pair of curves (or parentheses; see Sec. 16.6), the pair of dashes (see Sec. 16.5), and the pair of commas.

Most writers often use the pair of commas as a light parenthesis, although they may not think of the usage in these terms. A common fault is the omission of the second comma of a parenthetical pair. This confuses the reader, because he is momentarily uncertain where the parenthetical remark ends and the main stream resumes. He has had one pause indicated; he feels the need for a second, but he does not find it.

Original Example	*Suggested Revision*
The train passes the plants of several large industrial shippers, such as the Davison Chemical Co. and the American Oil Co. which fill many cars every day.	The train passes the plants of several large industrial shippers, such as the Davison Chemical Co. and the American Oil Co., which fill many cars every day. (Or omit *both* commas.)
The distance between Baltimore, Md. and Washington, D. C. is about 40 miles.	The distance between Baltimore, Md., and Washington, D. C., is about 40 miles.
The agreement ran from Jan. 1, 1970 to Dec. 31, 1972.	The agreement ran from Jan. 1, 1970, to Dec. 31, 1972.
Argentina, the second largest nation in South America with an area of over a million square miles, is located in the southeastern part of the continent.	Argentina, the second largest nation in South America, with an area of over a million square miles, is located in the southeastern part of the continent.

16.2.8 The Comma between Closely Related Parts

Since the comma is a separator, it should *not* go between elements that are very closely related. In particular, a comma should not separate a verb from its subject, its object, or its complement. The temptation to put a comma between subject and verb is strongest with a long subject that seems to call for a breather. A better solution is to rewrite the sentence or shorten the subject.

Original Example	*Suggested Revision*
That he runs from danger, is an important indicator of a man's character.	That he runs from danger is an important indicator of a man's character.

Notice, however, that a *pair* of commas between verb and subject is all right because the second comma of the pair shows that the interruption is over and the main stream has been rejoined.

The brave man, although afraid, does not run from danger.

16.3 The Semicolon

It is possible to write grammatically and clearly without ever using a semicolon. But the semicolon, as an intermediate-weight separator, can often help you express meaning clearly. The semicolon is most likely to be useful in three constructions:

16.3.1 The Semicolon with "Heavy" Connective Words

When certain so-called "heavy" or "uncommon" connective words (adverbs) join two independent clauses, the rules call for a stop heavier than a comma preceding them. A semicolon is suitable. Chief among this group of words are *however, therefore, nevertheless, moreover, otherwise, hence, also, thus,* and *yet.*

If you put a comma before any of these words when it is being used as a connective, you are guilty of a sin known as the *comma splice.* This fault leads to only very momentary confusion, but it is a mark of the unpracticed writer; therefore you should avoid it in order to maintain your reputation.

Original Example	*Suggested Revision*
The engine was badly damaged, nevertheless, the plane still flew.	The engine was badly damaged; nevertheless, the plane still flew.
The circuit is clearly shown, therefore, I shall not describe it in detail.	The circuit is clearly shown; therefore, I shall not describe it in detail.
The angle is not correct, however, it approximates the required deviation.	The angle is not correct; however, it approximates the required deviation.

(Any of these semicolons could be replaced by a period.)

Notice that the rule is corroborated by the voice-drop test: the break before these connective words is slightly greater than the one after them.

The words of this group are often used not as connectives *between* independent clauses but as adverbs *within* clauses. In this construction they may be enclosed in a pair of parenthetical commas.

> The engine was badly damaged. The plane, nevertheless, still flew.
>
> The angle is not correct. It does, however, approximate the required deviation.

16.3.2 The Semicolon between Closely Related Independent Clauses

The close logical relationship between two independent clauses (sentences) without any connective word can be indicated by a semicolon.

> I shall be out of town Tuesday; you may come to see me Wednesday.

This use of the semicolon can be helpful in avoiding a monotonous succession of short declarative sentences (Sec. 14.6).

Original Example	*Suggested Revision*
The gas inlet is a rectangular pipe entering tangential to the body of the cyclone. The gas outlet is an inner concentric cylinder extending down into the body of the cyclone. The outlet for deposited dust is at the bottom of the cone.	The gas inlet is a rectangular pipe entering tangential to the body of the cyclone; the gas outlet is an inner concentric cylinder extending down into the body. The outlet for deposited dust is at the bottom of the cone.

But if you use a comma to separate two independent clauses without any connective word, you will again be guilty of a comma splice. Don't do it.

Original Example	*Suggested Revision*
The transistor eliminated the problem of heating, it consumes only about a millionth the power of an equivalent vacuum tube.	The transistor eliminated the problem of heating, *because* it consumes only about a millionth the power of an equivalent vacuum tube. *or:* The transistor eliminated the problem of heating; it consumes only about a millionth the power of an equivalent vacuum tube. *or:* The transistor eliminated the problem of heating. It consumes only about a millionth the power of an equivalent vacuum tube.

16.3.3 The Semicolon Instead of Commas

Sentence elements that are normally separated by commas may themselves contain internal commas. It is then sometimes difficult to distinguish between the internal commas and the external commas that are supposed to separate the elements. Whenever commas are liable to become confused in this way, semicolons may be substituted at your discretion for the external, or heavier, commas.

Original Example	*Suggested Revision*
A large amount of freight and many passengers are moved in and out via TWA, United, North Central, and Lake Central Airlines, Grand Trunk, New York Central, Penn Central, Pere Marquette, and Chesapeake and Ohio Railroads, and U.S. 131 and 16 and Michigan 21, 37, 44, and 50.	A large amount of freight and many passengers are moved in and out via TWA, United, North Central, and Lake Central Airlines; Grand Trunk, New York Central, Penn Central, Pere Marquette, and Chesapeake and Ohio Railroads; and U.S. 131 and 16 and Michigan 21, 37, 44, and 50.
Such a program could have been initiated through written discussions in the leading farm journals, missionary salesmen to discuss the use, applications, and benefits of the fertilizer at the local Grange or other similar organizational meetings, and experiments conducted by the various large universities which have agricultural schools.	Such a program could have been initiated through written discussions in the leading farm journals; missionary salesmen to discuss the use, applications, and benefits of the fertilizer at the local Grange or other similar organizational meetings; and experiments conducted by the various large universities which have agricultural schools.

Of course either of these sentences might be further clarified by the addition of numbers for the items set off in semicolons, or perhaps by being broken up into several sentences.

Notice that in both revised sentences the conjunction *and* before the last member of the series is preceded by a semicolon (not a comma), in order to keep the punctuation parallel.

16.4 The Colon

16.4.1 General Function of the Colon

The colon was once a member of the series of pause-indicators that now consists of the period, the semicolon, and the comma. In modern technical writing the colon is a separate, highly specialized punctuation mark. Its function may be illustrated this way:

Promise: fulfillment.

That is, the colon comes after a statement that promises more information; it separates the promise from the fulfillment. Because the colon is reserved for this special function, it acts as a signal. Whenever you see a colon (except in certain conventional uses; see Sec. 16.4.5), you can expect further information about the subject that has just been mentioned.

16.4.2 The Colon to Introduce a Formal List

The simplest and commonest promise-fulfillment function is the introduction of a formal list.

> These units were chosen for the circuits: 7AK7, 7AD7, 3E29, 6YG6, and 6SN7.
>
> This work has included the measurement of the following parameters: (1) charging-current distribution; (2) spot interaction; (3) spot stability; and (4) gate lengths and amplitudes.
>
> The requirements for a satisfactory voltmeter for this service are:
> 1. It must be accurate to ±2 volts at full scale.
> 2. It must be insensitive to accelerations of 5 g.
> 3. It must be easily readable from 5 feet.

Because the colon is a further-information signal, you can usually omit the words *as follows,* or *the following,* from the introduction to a list. The colon itself leads the reader to expect more to follow.

16.4.3 The Colon to Introduce a Formal Quotation

The colon is used to introduce a formal quotation.

> The instruction manual has this to say: "With the left foot depress the clutch pedal. Move the gear-shift lever to the left and forward. With the right foot depress the accelerator pedal slightly. Then slowly release the clutch pedal."

16.4.4 The Colon to Introduce Further Explanation

In a less common construction, the colon introduces a statement that supplies *further explanation* of the statement that precedes it.

> Load conditions in the test must duplicate actual operating conditions: first, the capacitance of the ignition harness must be 200 micromicrofarads; second, the shunt resistance of the spark gap must be kept to a minimum.
>
> The damper control has a dual function: to maintain the proper proportion of cold air and heated air by regulating the mixing damper, and to control the velocity of the mixed air by regulating the outlet damper.

16.4.5 The Colon in Conventional Constructions

In addition, the colon is used in certain well-known conventional constructions, among them:

Between hours and minutes in an expression of time (7:28 A.M.)

After the greeting in a formal business letter (Gentlemen:)

Between various items in bibliographical entries (see Sec. 17.5.1)

16.5 The Dash

16.5.1 General Function of the Dash

The dash—another specialized punctuation mark—in general announces a discontinuity. (Do not confuse the dash with the hyphen, which is physically shorter and which serves an entirely different purpose; see Sec. 16.7.) As a signal of interruption, the dash is a useful but not essential mark whose functions overlap those of other marks. The rest of Sec. 16.5 describes several specific uses of the dash. Of these, the ones discussed in Secs. 16.5.2, 16.5.3, and 16.5.4 are appropriate in technical writing; the rest should usually be avoided.

16.5.2 Dashes Paired as Parentheses

A pair of dashes makes intermediate-weight parentheses (see Sec. 16.2.7). An example of this usage appears in the first sentence of Sec. 16.5.1; the parenthetical remark could have been set off in a pair of commas, but they would have provided rather inadequate pauses for the interruption, or discontinuity. Similarly, the parenthetical remark could be set off in a pair of curves (Sec. 16.6), but they would separate it a little too completely from *dash,* to which it is logically close. The pair of dashes in that sentence conveys a shade of meaning that would be hard to express otherwise.

If the second dash of a parenthetical pair comes at the end of the sentence, it is omitted.

Original Example	*Suggested Revision*
The four methods of testing samples in solution, by use of a macro scheme of analysis, a micro scheme, a semimicro scheme, or spot tests, are of the greatest interest.	The four methods of testing samples in solution—by use of a macro scheme of analysis, a micro scheme, a semimicro scheme, or spot tests—are of the greatest interest.

16.5.3 The Dash to Introduce a List

The dash may be used to introduce a list, like an informal colon (but do not use both a dash and a colon together).

> Some of the chassis members contribute to both unsprung weight and sprung weight—the driveshaft, the suspension links, and the springs themselves.

The dash is often used to avoid confusion in a series. Take this example:

> He had put up for sale three pieces of machinery—a lathe, an electric saw, and a drill.

Here the author intended that three objects should be noted, not six. A comma might have caused such confusion; on the other hand, a colon could have been used.

16.5.4 The Dash to Introduce Further Explanation

Again like an informal colon, the dash may be used to announce further explanation of the statement that precedes it.

> In automatic service the engine undergoes fluctuating loads—both its torque and its speed are constantly changing.
>
> Sometimes we want a simplified description—a description that even a layman can understand easily.

16.5.5 The Dash in Other Uses

In nontechnical writing the dash is legitimately used for a number of other specialized purposes, such as to provide a pause before an unexpected word or a supposedly witty remark. Even in the most informal writing these uses of the dash should be employed sparingly if they are to be effective. In technical writing they should not be employed at all.

Some inexperienced or lazy writers use a great many dashes indiscriminately in place of other punctuation marks. This practice is, of course, confusing to the reader.

16.6 The Parenthesis

The general function of parentheses (or curves) is so well understood that no discussion is needed here.

16.6.1 The Parenthesis and Other Punctuation

The parenthesis should not be preceded by a comma. If a comma is required after the sentence element that comes before the parenthetical remark, *the comma should follow the parenthetical remark*. This rule is sensible because the parenthetical remark is usually logically connected to what has gone before it. For example:

> Although the engineer was thoroughly familiar with the crossing (he had made this run daily for three years), he failed to slow down in time.

Notice that if the parenthetical remark were removed, the comma would come after *crossing.*

When a parenthetical remark is included as part of a sentence (even at the end of the sentence), *the period goes outside the second curve,* and the parenthetical remark does not begin with a capital, even though it is a complete sentence (see example above). When a whole sentence stands alone in parentheses, the period goes inside the second curve, and the sentence begins with a capital. (This might be a good place to point out that the overuse of parentheses is often a sign of incomplete organization.)

16.6.2 Parentheses around Equation Numbers

Parentheses are customarily used around the numerals that distinguish items in an enumeration; for examples, see Secs. 16.2.4 and 16.4.2.

A special case of this usage is the numbering of equations. In any treatise or derivation that contains more than just a few equations, it is customary to number the equations serially. Each equation number is enclosed in parentheses and is usually placed at the right-hand margin opposite the equation.

This use of parentheses is logical and convenient. But many writers carry the parentheses over into their references to the equation numbers:

> Values of these parameters may be obtained by making the proper substitutions in Equations (11), (13), and (17).

In this construction the parentheses are separating the inseparable. The name of a certain equation is Equation 11, and there is no more reason to put *11* in parentheses than *Equation.* Whenever you refer to an equation (or other numbered item) by number, omit the parentheses even though they appear in the actual enumeration.

> . . . substitutions in Equations 11, 13, and 17.

16.6.3 Square Brackets

Square brackets (as distinguished from the usual curved parentheses) have a specialized function. They are put around material interpolated into a quotation. They indicate that the bracketed words or symbols are not part of the quoted passage, but are remarks or further explanation added by the quoter or editor.

> The article stated, "When this cycling control [a two-position control for Ward-Leonard dynamometers] is in operation, its resistors replace the usual manually operated field rheostat."

16.7 The Hyphen

16.7.1 The Hyphen in Compound Nouns

A noun consisting of two or more words that name one object is called a compound noun. Some compound nouns are written as separate words; some are hyphenated; and more and more of them are being fused into single words. The only way to tell how any particular compound noun should be written is to look it up in your dictionary or some other reference. Even then, you will find that authorities differ about many words. Here are a few examples of compound nouns.

Separate Words	*Hyphenated*	*Fused*
name plate	flare-up	setup (noun)
center punch	light-year	centerboard
center of gravity	lean-to	seaplane
sea otter	double-header	strikebreaker

16.7.2 The Hyphen in Compound Adjectives

An adjective consisting of two or more words that express a unified idea —for instance, *red-nosed*—is called a compound adjective. The words that go to make up a compound adjective may themselves be adjectives, nouns, adverbs, or any other parts of speech.

A compound adjective that comes before the noun it modifies should be hyphenated. Although this rule is commonly relaxed, it should be followed quite rigorously in technical writing. Technical subjects seem to call for a lot of compound adjectives, some of them quite complex; and technical writing must be precise. If you do not hyphenate compound adjectives, your reader will often be momentarily delayed, sometimes actually confused.

Consider this passage:

> . . . as given for the adiabatic, no shaft work, constant gravitation potential system.

The reader who is a specialist in thermodynamics quickly spots the compounds. The nonspecialist comes upon these groups of words and discovers at the end of each that it is a compound. This process of comprehension requires unnecessary mental backtracking. The difficulty is aggravated if the compound is separated by the ending of a line. On the other hand, the compounds are immediately evident when they are hyphenated:

> . . . as given for the adiabatic, no-shaft-work, constant-gravitation-potential system.

The following examples illustrate some of the temporary misunderstandings that can result from unhyphenated compound adjectives. When you read them, notice that the need for the hyphen can usually be indicated by the reading-aloud test: the words that should be hyphenated are partially fused together as you say them.

Original Example	*Comment*	*Suggested Revision*
curved nose cutting tool	For a moment you might read *curved nose-cutting tool.*	curved-nose cutting tool
big city office	A big office in the city, or an office in the big city?	big-city office
carbon containing compounds	Carbon that contains compounds?	carbon-containing compounds
The contractor obtained good results with eight cubic yard scrapers.	Eight scrapers each of 1 cubic yard?	The contractor obtained good results with eight-cubic-yard scrapers.
small time delay transformer	A small transformer that provides time delay?	small-time-delay transformer

Here is a list of a few representative phrases containing compound adjectives properly hyphenated.

110-volt line
30-foot depth
160-horsepower engine
signal-to-noise ratio
steady-state operation
single-phase motor
high-fidelity system
frequency-modulation station
cathode-ray tube
line-of-sight transmission
short-wave band
high-frequency system
long-distance transmission
information-handling system
cable-operated brakes
spark-plug wrench
clutch-pedal linkage

Notice that the same words when used as nouns do not require hyphens:

110 volts
30 feet
operation in the steady state
The system has high fidelity.
a cathode ray
The system operates at high frequency.
Power is transmitted over a long distance.
a spark plug
the clutch pedal

The hyphen can be conveniently used as a signal to show that a word will be connected to another word later in the sentence to form a compound adjective.

two-, three-, and four-digit numbers

Exceptions: The hyphen is usually omitted from compound adjectives:

(1) When the first word is an adverb ending in -ly, which obviously is going to operate on the word that follows it:

a partly digested fish.

(2) When the compound adjective is a proper noun:

the United States flag.

(3) When the compound adjective is the name of a chemical compound:

a copper sulfate solution.

(4) When the compound adjective is made from a commonly used two-word noun:

an oil filter salesman.

16.8 The Apostrophe

If we are to judge by much of the writing coming from both professional men and students, the apostrophe is rapidly going into the discard. This discussion, then, is included only as a word of caution. There are still places where the apostrophe should be used—for example, when one speaks of *my brother's son* or *the company's proposals.*

Of course the use of the apostrophe can become awkward, and possession can be shown by an indirect expression such as: *the design of the machine tools* instead of *the machine tools' design.*

A point of confusion occurs when the apostrophe is dropped and the possessive is used as an adjective, as when *the company's proposals* is changed to *the company proposals.*

In short, we say that if you do not use an apostrophe to show possession, be prepared to defend your choice.

RECOMMENDED READING

Now that you have studied the principles of technical writing in Chapters 14, 15, and 16, you will be able to appreciate an amusing and instructive little book called *The Elements of Style,* by William Strunk, Jr., and E. B. White (Macmillan).

17. Mechanics

Every student dreams about the day when he will have a secretary who will attend to all the tiresome mechanical details that come up whenever he has to write a report or paper. This chapter contains advice about a lot of matters that you might think are her business, not yours. But the chances are that when you finally acquire the services of a secretary, she will have been trained at a school that has never heard of technical reporting. She will probably be more interested in various niceties of layout than she will in extreme clarity.

We intend no slight to secretarial schools or their products. But you will find that most secretaries require close supervision and detailed instructions on the subjects treated in this chapter.

17.1 Abbreviations

17.1.1 Use of Abbreviations

An abbreviation will not communicate information to your readers unless it is familiar to them. Thus you should fit your use of abbreviations to the understanding of your intended readers. If in doubt, don't abbreviate. If you have occasion to use an abbreviation repeatedly and you question whether your readers are familiar with it, you can explain it parenthetically the first time it appears.

The propulsion units in ships of this class seldom produce more than 10,000 shp (shaft horsepower).

Abbreviations should be used sparingly in text matter. There are two exceptions to this generally observed rule, and it would be pedantic to insist that terms in either of these groups be spelled out for a technical audience:

(1) Names of units *preceded by numerals;* for instance:

an attenuation of 10 db
a 120-hp engine
The motor was running at 3600 rpm.

But

The decibel is a logarithmic unit.
a fractional-horsepower motor
Motor speed is measured in revolutions per minute.

(2) Terms whose abbreviations have become so well known that they are commonly accepted as part of our everyday vocabulary; for instance:

a-c, d-c, emf, rms, Fig., bmep

On graphs, diagrams, or tables where space is at a premium, abbreviations may be used freely, provided their meaning is clear.

17.1.2 Form of Abbreviations

When it comes to the form that abbreviations will take, chaos reigns. You will often find the same term abbreviated in several different ways in various well-edited publications. But scientists and engineers are more and more falling in line with the American National Standards Institute's *Abbreviations for Scientific and Engineering Terms* (Z10.1—1941). This pamphlet advocates the following general rules for scientific and technical abbreviations:

(1) Omit periods unless the abbreviation happens to spell a word or would otherwise be confusing without the period.

(2) Do not use capitals unless the word abbreviated is a proper noun.

(3) Use the same abbreviation for singular and plural.

Thus:

cm = centimeter or centimeters
in. = inch or inches
Btu = British thermal unit

The American National Standards Institute's list of abbreviations is reproduced in the Appendix, Sec. 21.1.

Symbols such as # for *pound* or *number,* ' for *foot,* " for *inch,* / for *per* are not used in the text sections of some publications; they may be used freely in graphs or tables if there is not room for the corresponding abbreviations.

Exceptions: Two symbols that are commonly considered permissible in text are $ for *dollar* and % for *percent.*

Numerical expressions of temperature usually include a letter to designate the scale, but the degree symbol and the period are preferably omitted.

> Water boils at 212 F (100 C).

17.2 Numerals

17.2.1 Use of Numerals

In high school you may have been taught that numbers below 10 should be spelled out, numbers of 10 or higher expressed in Arabic numerals (hereafter called simply *numerals*). But, depending on where you went to school, the dividing line may not have been 10: it may have been 13, or 11, or 3; or perhaps you were told to spell out numbers that can be expressed in one word, or in two words, or in three words. Blind adherence to one of these arbitrary rules might produce:

> Radio receiving sets commonly contain anywhere from four to 24 transistors.

The inconsistent expression of these numbers actually obscures the comparison between 4 transistors and 24 transistors. Even the purists will usually say that where numbers below the dividing line (whatever arbitrary line happens to be chosen) are mixed with numbers above it, you may use numerals for all of them.

Now, what is the size of the unit in which a mixture of numbers permits us to use numerals for all of them? A sentence? A paragraph? A page? The authorities do not say.

But we can draw an inference from this widely accepted rule: *In technical and business writing, where there are frequent expressions of quantity, numerals are generally more acceptable than they are in social or literary writing. Numerals may be used freely to express numbers of any size unless they would cause confusion or unless a written-out number seems more appropriate* (see Sec. 17.2.3).

17.2.2 Numerals Specifically Acceptable

Numerals are accepted everywhere (except in wedding invitations) for the following kinds of expressions:

(1) Exact quantities when the number goes with units of measurement.

a 12-volt storage battery
The standard size is 8½ by 11 inches.
a 3-lb hammer
The temperature fell to 5 F.
the 60-cycle 110-volt line
The bond yields 6 percent.
The fine was $5.

(2) Decimals, fractions, mixed numbers.

The specific heat of water is 1.0.
Water is about 1⅓ times as heavy as gasoline.

(3) Numbered objects in a list.

Fig. 2
column 3
page 26
Chapter 6
Equation 4
No. 7

(4) Dates, addresses, and time of day.

The elevator in the building at 26 Prince St. fell at 2:35 P.M., July 6, 1971.

17.2.3 Numerals Not Acceptable

Numerals are usually not used:

(1) At the beginning of a sentence. If the number is not too long or too complicated, it can be spelled out; if this is not feasible, the sentence should be rewritten so that the number is not at the beginning.

(2) Adjacent to other numerals. When two numbers are adjacent and not separated by punctuation, there is danger of misreading (or misprinting) unless one of them is written out.

Original Example	*Suggested Revision*
24 5-inch straps	twenty-four 5-inch straps *or:* 24 five-inch straps

(3) In round-number approximations. Numerals produce an air of precision that is not suitable for approximate quantities, which are better spelled out, or partially spelled out as in the second example below.

I expect to be in your territory for a week or two.
The deficit this year will be about $15 billion.

Numbers rounded off to even millions or billions are often easier to comprehend when the millions or billions are spelled out. This device should not be used for numbers smaller than a million.

It often seems appropriate to spell out very small numbers—particularly when they refer to objects or persons rather than units.

This paper synthesizes the findings of three independent investigators.

17.2.4 Numerals: Miscellaneous

(1) Decimal numbers of less than unity should be written with a zero preceding the decimal point. This practice helps to prevent typists from omitting the decimal point, as well as reducing the likelihood of misreading.

The odometer advances every 0.1 mile.

(2) Numbers should *not* be expressed in both words and numerals, except perhaps in legal documents. Let us leave these to the lawyers. Do *not* write:

The trusses will be ready in sixty (60) days.

(3) When you wish to express a range of numbers, use a complete idiom.

Incomplete or Mixed Idiom	*Complete Idiom*
Delivery is promised in from 30–60 days.	Delivery is promised in from 30 *to* 60 days. *or:* Delivery is promised in 30–60 days.
The work will require between 30–60 days.	The work will require between 30 *and* 60 days. *or:* The work will require 30–60 days.

(4) References to numbered items do not need the word *Number*. Its omission is particularly helpful in the plural form.

Original Example	*Suggested Revision*
See Drawing No. A-30340.	See Drawing A-30340.
. . . as described in Report No. GP-127.	. . . as described in Report GP-127.
Reports No. 5 and 6 Reports Nos. 5 and 6	Reports 5 and 6

17.3 Capitals

Initial capital letters used according to conventions with which you are entirely familiar convey certain information. In addition, they give emphasis to the words they begin.

Some technical writers tend to use initial capitals too freely, especially for the names of devices and components. This practice is likely to give undue emphasis to some words. At the same time, if capitals are overused they lose their force and their significance; their ability to emphasize is weakened, and their conventional meanings are beclouded. In general, capitalization should be kept to a minimum.

Original Example	*Suggested Revision*
The major components of the High Speed Computer are the Arithmetic Element, the Control, the Storage Element, and the Input Output System.	The major components of the high-speed computer are the arithmetic element, the control, the storage element, and the input-output system.

Notice that proper hyphenation of the compound adjectives (see Sec. 16.7.2) performs one of the functions that the original capitalization was probably intended for: to make it clear that certain long names are actually integrated units.

Most, but not all, publications capitalize specific numbered items such as Figure 2 (or Fig. 2) and Table VI. When they occur in text, these expressions are usually repetitions of captions that have appeared capitalized under (or over) the figures or tables that they describe. Thus it seems consistent to capitalize them in the text, simply repeating the titles literally. (The same criterion can regulate the choice between *Fig.* and *Figure* in text: simply reproduce the title that appears under the figure, whether it is Fig. 2—which is entirely permissible—or Figure 2.)

17.4 Quotation Marks; Underlining (Italics)

17.4.1 Mechanics

When a single quotation runs more than one paragraph, quotation marks (often called *quotes*) are put at the beginning of each paragraph, but at the end of only the last one.

A quotation within a quotation is set off with single marks, as in the second example below.

A convention dictates that a comma or a period contiguous to a quotation mark shall always be *inside* the quotation, whether it is logically part of the quotation or part of the main sentence.

"It is the duty of every engineer," said the dean, "to consider the social consequences of the machines and devices he designs and builds."

But other punctuation marks go either inside or outside the quotation depending on whether they are logically part of the quoted passage or of the main sentence.

The dean raised an important question: "What," he asked, "do you mean when you say, 'The curriculum is no longer relevant'?" But does the dean have the right to tell his students "You must not do this and this and this"?

For some of the functions of quotation marks, underlining (italics in printed matter) is often used instead; see Secs. 17.4.3 and 17.4.5.

17.4.2 Direct Quotations

Whenever you quote somebody else's words, you of course owe it to him to put quotation marks around them and to give him credit for them, no matter how short the passage.

An equally acceptable device is to indent the quoted material inside the margins of the main text and omit the quotation marks. In double-spaced typewritten manuscript, indented quotations are usually single-spaced. In printed material, they are usually set in type smaller than that used for the main text.

17.4.3 Quotation Marks to Distinguish

Quotation marks are often used—and overused—to set aside or distinguish words or phrases. This usage seems to fall into three main categories:

(1) To indicate that a word or phrase is being employed in a new or special or unusual sense, or to introduce a new or strange term. Underlining is sometimes used for the same purpose. The new or strange word should be in quotes (or underlined or italicized) only at its first appearance. From then on, unless confusion would result, it should be considered well enough established to stand on its own feet without special indication.

Any violation of these arbitrary rules has marked a member of our society as "improper." By this criterion, many serious students must be classified as improper.

(2) Around slangy or slightly disreputable expressions. Quotation marks around slightly off-color expressions serve as an apology. They say, for the writer, "I know better than this, but if you will allow me. . . ." You should probably never use quotation marks this way. Decide beforehand whether the expression is suitable for your intended audience. If it is, use it without apology and without quotation marks. If it is not, use a different expression.

(3) To emphasize a word or phrase, or to set it apart. Technical writers sometimes put quotation marks around such terms as the names of parts of a system, or the words from a nameplate or panel. This indiscriminate use of quotation marks serves to rob them of meaning and to reduce their effectiveness when they are used properly. Often the quotation marks can simply be omitted from such constructions. If some sort of emphasis or separation is required, underlining (italics) is likely to be more suitable than quotes.

Original Example	*Suggested Revision*
The "accumulator" is connected to the supply.	The accumulator is connected to the supply.
Turn the switch to the "on" position.	Turn the switch to the *on* position.
The first digit column contains a "one."	The first digit column contains a 1.

17.4.4 Quotation Marks in Bibliographical Entries

In formal bibliographical entries, titles of books and periodicals are customarily underlined (italicized), while titles of chapters and articles are enclosed in quotation marks (see Sec. 17.5). Some people use quotation marks instead of underlining for the titles of books and periodicals.

17.4.5 Words Referred to as Words

When you wish to mention a word as a word—as a vocabulary term—rather than to refer to its meaning, you can either underline (italicize) the word or enclose it in quotes.

My wife thinks *stink* is an ugly word.
(or . . . "stink" . . .)

17.5 Bibliographies

17.5.1 Form of Bibliographical Entries

Technical reports and papers frequently carry bibliographies. If you have looked critically at the bibliographies in various publications, you have discovered that there is little uniformity from one publication to another. If you are writing for any particular publication, you will do well to follow its style.

But in general a bibliographical entry is successful if it enables the reader to find the cited material with a minimum of trouble; the order of the parts and the punctuation used to separate them is relatively unimportant. The required information—conventionally given in this order—usually is:

Books	*Magazine and Journal Articles*
Author	Author
Title	Title of article
(Place of publication)	Name of magazine or journal
Publisher	Volume number, date of issue, or both
Date	Pages

Encyclopedia Articles	*Reports*
Title of article	Author
Name of encyclopedia	Title
Edition number	Name of issuing organization
Date	Report number
Volume number	Date
Pages	

Here are examples of acceptable bibliographical entries:

(*1*) *Book*

Hogg, R. V., and Craig, A. T. *Introduction to Mathematical Statistics.* New York: The Macmillan Co., 1966.

(*2*) *Magazine and journal articles*

Lear, L. "Northwest Passage to What?" *Saturday Review* (Nov. 1, 1969), 55–67.

Cassini, M. A. "Profile of an American Industry," *Steelways,* Vol. 25, No. 3 (1969), 2–8.

(*3*) *Encyclopedia article*

"Vacuum Pumps." *Encyclopedia Britannica* (1968), Vol. 22, 838.

(*4*) *Report*

Moore, V. S., Bretnall, W. D., and Stetson, A. R. *Evaluation of Coatings for Cobalt- and Nickel-Base Superalloys.* Solar Division of International Harvester Company Report RDR 1474-2.

Anonymous articles from magazines or other publications are treated in the same way as the encyclopedia example above, with the title first.

Some publications list joint authors this way:

Jones, P. D., H. R. Smith, and R. D. Leffingwell.

The different order for surnames and initials seems to be a needless inconsistency; we recommend instead:

Jones, P. D., Smith, H. R., and Leffingwell, R. D.

Titles of books, magazines, and articles are customarily underlined or put in quotation marks to distinguish them from surrounding text (see Sec. 17.4). In a formal bibliography, one expects to see titles, and they hardly need to be distinguished by special typography. Some publications, particularly typewritten reports, simply ignore the underlining convention in their bibliographies.

Whatever bibliographical form you choose, use it consistently for any one document. For sample bibliographies see pages 120 and 406.

17.5.2 Arrangement of Bibliographies

The entries in a bibliography may be arranged in various ways. The simplest and most usual is a single list, with authors' surnames and titles of anonymous articles intermixed in alphabetical sequence.

Longer bibliographies are sometimes broken up into separate lists of books, articles, pamphlets, or other forms of publication. Then again, it may be helpful to your readers to break your bibliography into separate lists by topics in a long report or paper.

Another acceptable sequence for bibliographical entries is the order in which you refer to them. This order is appropriate particularly if you use the bibliography instead of individual footnotes or citations (see Sec. 17.6.2).

17.6 Footnotes (Citation)

17.6.1 What to Credit

Remember that you are obligated to give credit not only for sequences of *words* that you have borrowed, but also for sequences, or essential arrangements, of *ideas*. If you particularly want your readers to know who the original author was, or if his name will lend weight and authority to your statement, you can make your citation right in the text. Usually, though, the citation is a routine matter, and a footnote is a convenient way to take care of it.

Authors of papers based largely on library research often wonder how frequent their footnotes must be. Must every sentence have a footnote giving the source of its information? Or is one general statement of indebtedness sufficient? The answer will usually be a compromise between these two extremes. If you follow the rule below, you will probably fulfill your obligations without cluttering your manuscript with needless footnotes.

To give credit for an arrangement of ideas—that is, for a considerable mass of information—use one general, overall citation. To give credit for a specific arrangement of words or lines—that is, for a verbatim quotation or for the reproduction of a drawing—provide a citation for each item.

Information that is common knowledge—that you might find in any of several sources—does not need to be credited unless you present it in a verbatim quotation.

17.6.2 Systems for Citation

Two systems for citing references are in common use. A third, which is a sort of compromise between the first two, has advantages that are making it increasingly popular.

(1) *Full bibliographical notes.* Each superscript in the text refers to a note, which may be at the foot of the page or in a group at the end of the chapter or the end of the paper (but separate from the bibliography). The note for the first reference to each source is a full bibliographical entry, with the addition of specific page numbers. (The order of the items in notes often differs from the conventional order in bibliographies: the title may precede the author's name, and the author's given name or initials may precede his surname.) Succeeding notes referring to the same source may be abbreviated in any of the several ways described in most handbooks. The terms *ibid.* (in the same place) and *loc. cit.* (in the place already mentioned) may be used if there is no doubt about what they refer to. Perhaps better, the bibliographical entry may be shortened down to the point where it carries just enough information to identify its reference: if only one work of an author is referred to, his surname alone may be sufficient; or maybe a shortened form of the title, containing only key words, must be added.

This system is the conventional one in most of the learned journals. It is complete, explicit, and clear. But it is cumbersome and it involves a lot of duplication; and if the notes are put at the foot of each page, they produce a cluttered appearance.

(2) *Bibliography instead of notes.* Each superscript in the text refers to the similarly numbered entry in the bibliography. This system, which is used by a number of the technical journals, is neat and convenient, but it does not provide for the listing of specific page numbers for each reference.

(3) *Footnotes referring to bibliography.* Each superscript refers to a note (at the foot of the page or grouped) that cites a bibliography entry *by number* and gives a specific page number. For example:

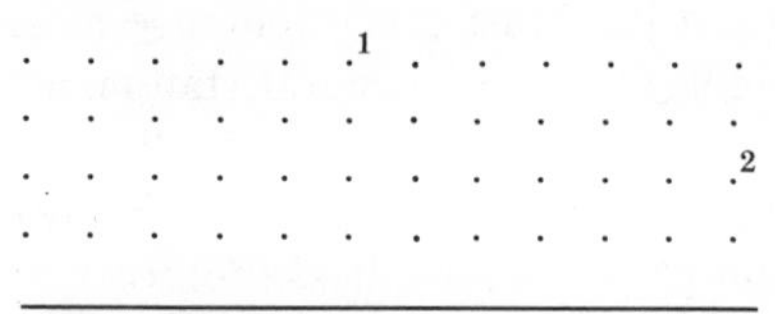

[1] Reference 6, p. 29.

[2] Reference 14, pp. 142–150.

This system is the same as No. 1 except that bibliography numbers are used instead of titles. It is less cumbersome than No. 1; yet it permits references to specific page numbers.

If you choose a system that does not use actual footnotes, place one footnote at the bottom of the page carrying the first superscript, and in this footnote explain the system you are using and tell your readers where to look for the references. For instance, you might say

> [1] Superscripts refer to similarly numbered entries in the Bibliography.

or

> [1] Superscripts refer to notes at the end of each chapter.

This explanation will prevent your readers from referring to the bibliography when they should be referring to notes, or from looking for notes that are not there.

Superscript numerals are generally preferable to asterisks, daggers, or other symbols for referring to notes, particularly when there are several references on one page. A new series starting with 1 may be used for each page or each chapter, or a single series may be used for the entire paper or report.

17.7 Heading Systems

Section 2.8 stressed the idea that headings and subheadings should be used freely, and that they should make evident the structure of the paper or report. Therefore it is very important to employ a system of headings in which the relative weights of the various headings, and their relationships to each other, are abundantly clear. Three effective systems are described in Secs. 17.7.1–17.7.3.

Of these schemes, the first two are customarily used for outlines as well as text. The third—a less formal one—is suitable for text but not for outlines.

The examples that follow are set up with acceptable arrangements of typography and indentation. But of course many other arrangements are entirely suitable for the same basic schemes. Just be sure that the typography and indentation you choose contribute to the clarity and information-content of your heading system.

17.7.1 The Decimal System

The system of headings used in this book is commonly called the *decimal system,* although the periods are really simply separators, not decimal points. The numbers and headings are arranged this way:

1. FIRST-DEGREE HEADING
    1.1 Second-degree Heading
    1.2 Second-degree Heading
        1.2.1 Third-degree Heading

1.2.2 Third-degree Heading
1.3 Second-degree Heading

2. FIRST-DEGREE HEADING
Etc.

The number of degrees may be extended at will, of course, by the addition of more digits and more points. If you use this system for an outline, you may want six or more degrees, but for headings in a report or paper, three or four are usually enough.

The decimal system is becoming increasingly popular. It is a rigorous system that never leaves any doubt about the relative weight of any heading. Furthermore, when you come upon an isolated decimal-system subheading—no matter of what degree—you know at once how it fits into the main scheme.

Some users of the decimal system put a point only after the first-degree digit, running all lighter-weight digits together, this way:

1.
1.1
1.2
1.21
1.22
1.221
1.222
1.23
1.3
2.

This scheme is all right so long as there are not more than nine headings of the same degree in one group. But if there should be more than nine, ambiguity occurs: the heading 1.12, for instance, might then belong in either of two series:

1. 1.1 1.2 . . . 1.9 1.10 1.11 1.12	or	1. 1.1 1.11 1.12

Thus it is generally safer to add a point for each additional degree after the second; and it is essential whenever there are more than nine headings of the same degree in any one group.

A report using the decimal system of headings is reproduced in Sec. 21.2.15.

17.7.2 The Numeral-Letter System

Probably the commonest system of designations for outline headings uses a sequence of Roman numerals, capital letters, Arabic numerals, and lower-case letters.

I. FIRST-DEGREE HEADING
  A. Second-degree Heading
  B. Second-degree Heading
    1. Third-degree Heading
    2. Third-degree Heading
      a. Fourth-degree Heading
      b. Fourth-degree Heading
    3. Third-degree Heading
  C. Second-degree Heading

II. FIRST-DEGREE HEADING
  Etc.

The number of degrees may be extended by the addition of further Arabic numerals and lower-case letters in parentheses.

This system suffers the disadvantage that isolated subheadings are not oriented in the main scheme. When you come upon a heading numbered 3, for instance, you do not know whether 3 belongs under A, B, or C, etc.; and you do not know what Roman numeral the capital letter falls under. To find out, you must go backward until you come to a first-degree heading.

Specimen reports using numeral-letter headings will be found in Secs. 21.2.5 and 21.2.16.

17.7.3 Typography-Indentation Systems

The relative weights of various headings may be indicated by a consistent arrangement of typography and indentation. Such systems are not so rigorous as the ones described in the two preceding sections, and they are hardly suitable for outlines. But they are quite effective for text headings, particularly for relatively short and informal reports or papers.

One possible arrangement is shown at the top of the next page. Systems similar to this one are used in the Preface of this book and in some of the specimen reports in the Appendix. Of course many other variations are acceptable.

FIRST-DEGREE HEADING
(Centered; all capitals; underlined.)

SECOND-DEGREE HEADING
(At left margin; all capitals; underlined.)

Third-degree Heading
(Indented; initial capitals; not underlined.)

Examples of typography-indentation headings are illustrated in the specimen reports of Secs. 21.2.4 and 21.2.17.

17.7.4 Physical Arrangement of Headings

No matter what system of headings you use, you can improve its effectiveness by careful attention to physical arrangement.

(1) Headings of all degrees serve as signposts. They are effective only if they stand out from the surrounding text. To make them conspicuous:

(a) Put every heading, no matter of what degree, on a line by itself. Do not start a sentence on the line occupied by a heading.

(b) *Separate every heading or subheading from following text by two typewriter spaces (lines); separate every heading or subheading from preceding text by at least three typewriter spaces.* This spacing not only makes the heading stand out; it also ties the heading to the material with which it is logically connected and separates it from the preceding material.

(2) In reports or documents of any length, main (first-degree) headings usually go at the top of a new page.

(3) In outlines, the "text" (explanatory material) is usually indented as far as the heading under which it falls. Although the headings you use in a report or paper may be indented just as if they were in an outline, the text that comes under them should usually be carried to the left-hand margin (except, of course, for the first lines of paragraphs) no matter how far the heading is indented. Indentation of text under headings wastes paper and tends to spread things out unduly.

17.8 Typing

17.8.1 Line Spacing

Manuscripts for publication are always typed double-spaced. Business letters are customarily single-spaced. Reports may be either double-spaced or single-spaced, depending on the copy. Any material that contains equations, superscripts, or subscripts is easier to read if it is double-spaced. Reports or papers submitted to your instructors should be double-spaced to facilitate marking and correction.

17.8.2 Paragraph Indentation

In double-spaced copy the first line of each paragraph is customarily indented, say, five letter spaces. Paragraphs may be separated by a triple space.

In single-spaced copy a double space is left between paragraphs. In block-style single-spaced typing—popular for business letters—the paragraph is not indented, the double space serving as the only indication of a new paragraph. If one sentence ends at the bottom of a page and another one begins at the top of the next, there is no way for the reader to tell whether this second sentence begins a new paragraph or is part of the old one. Therefore in technical writing, where precision is more important than style, the first line of every paragraph should be indented whether the copy is double-spaced or single-spaced.

17.8.3 Hyphenation

Some words must be divided at the ends of lines to keep the right-hand margin from looking ragged. But try to keep the hyphenation at the ends of lines to a minimum.

17.9 Binding Margin

Most organizations bind their reports—at least the longer, more formal ones—in cardboard covers that conceal or obliterate at least ⅞ inch of the left-hand edge of the page. This concealed edge may be considerably wider in the interior of thick reports if their pages cannot be turned back sharply.

Therefore when you are preparing sheets of any kind—text, graphs, or drawings—that are to be bound into ordinary report covers, leave a binding margin of at least 1¼ inches, and preferably 1½ inches, at the left edge. Be particularly careful about graphs drawn on commercial graph paper, which often has a left-hand border of less than ¾ inch. We have seen curves plotted across the whole grid, right out to the border, with the scale in the border. When a curve-sheet like this is bound into a report, the vertical scale is completely hidden from view. To avoid this serious fault, you can either trace the graph on plain paper with a suitable margin, or you can get commercial graph paper with extra-wide borders (see Sec. 19.5).

17.10 Page Numbering

Two systems are in use for the numbering of pages in books and reports. In the more common of these, the preliminary pages (title page, preface, table of contents, and so forth) are numbered in one series of small Roman numerals; the main body is numbered in a separate series of Arabic numerals. In the other system, which is used occasionally for books and oftener for reports, all the pages are numbered consecutively in one series of Arabic numerals. The single series is in some ways more convenient for the reader.

Whichever system you use, the title page is counted as the first page (i or 1), although no number is placed on it.

On all ordinary pages the number usually goes at the *top outer corner.* The number is often omitted from pages that start with a centered main heading, but it is preferable, for the convenience of the reader, to put numbers at the *bottom center* of these pages.

17.11 Methods of Reproduction

This is not the place for an extended treatise on printing processes, but you should be interested in some general information about the various ways in which reports and other documents can be reproduced.

17.11.1 Copying Processes

(*1*) *Electrostatic Processes*

The most widely used office duplicating processes—known as xerography—are based on the attraction to an electrostatically charged surface of oppositely charged particles of pigment. The first of these electrostatic processes was trade-named Xerox, a term that is often used generically.

(*a*) *Xerox.* The Xerox process makes clear, high-grade copies of text or drawings and somewhat less clear copies of halftones. The largest machines, which are very expensive, are capable of producing several thousand copies per hour.

If the machine is kept busy, the cost per copy is relatively low; for small quantities, the cost is high. Frequently, however, the high cost is offset by the high quality of reproduction and the ease with which copies can be made, and Xerox is very popular. It has the advantage of using ordinary, untreated bond paper.

In the Xerox process, an image of the original is projected through an optical system onto a selenium-coated drum that has been given a positive electrostatic charge. The drum surface loses its charge wherever the light strikes (that is, in the white portions of the original). The drum is then sprinkled with negatively charged powdered pigment, which sticks only to the charged areas (the image). A positively charged sheet of paper is then rolled over the drum; it picks up the powder image, which is then fused on by an exposure to heat.

(*b*) *Liquid* or *Toner Electrostatic Process.* This electrostatic method requires a specially double-coated paper. Copy is clear but not so black as Xerox, and the special paper is rather limp. Copies have a high degree of permanence, but the coating cracks if strongly creased.

In this process, the coated paper is negatively charged and the image of the original projected directly onto it. The charge remains on the image portion, bleeds off from the white portion. The charged sheet is then passed through a liquid hydrocarbon toner in which positively charged pigment particles are suspended. The particles are attracted to the negatively charged image areas and fixed there by the toner. The volatile toner is then squeegeed and evaporated from the copy.

The machines used in this process are smaller and less expensive than Xerox.

(2) *Diazo* (*Ozalid*)

Drawings, illustrations, and text can be reproduced in small quantities by the diazo process, which is closely akin to photographic printing. The best-known representative of the diazo process (also known as black-and-white, blue-line, black-line, white-print, and ammonia-vapor process) is trade-named Ozalid.

In this process, a direct contact print is made from the original copy (which must be on translucent paper) onto photosensitive paper, which is then developed in ammonia vapor. The resulting copies are positive; they fade slowly over a period of years.

The process is highly flexible and very convenient. It is now on the wane, being used primarily for the duplication of large drawings or graphs, but it can also be used for text. *Typewritten copy that is to be reproduced by Ozalid should be carbon backed* (*typed with a sheet of carbon paper facing the back of the page*) *on translucent paper with no watermark.*

When more than fifty or a hundred copies of a line drawing are required, photo-offset printing costs no more and is of better quality than Ozalid. For text pages, Ditto, Mimeograph, or direct-image offset is cheaper when more than about half a dozen copies are required.

(3) *Dual-spectrum*

Small quantities of reasonably good copies can be made with simple equipment by the dual-spectrum method. Although the equipment is inexpensive, the unit cost of the copy paper and the intermediate sheets is relatively high, and the process is slow and cumbersome.

The dual-spectrum process receives its name from the two steps in the method. First, the original is exposed to light through a waxy photosensitive sheet, which picks up a latent image. Second, the intermediate sheet is heated in contact with a specially coated paper, which reproduces the image in black.

(4) *Thermofax*

The Thermofax process seems to be on the wane, but a number of machines are still in use, the process is quick and convenient, and it is useful for preparing spirit masters (see Sec. 17.11.2) and transparencies for the overhead projector (see Sec. 13.1.2).

In the Thermofax process, a waxy and rather brittle heat-sensitive paper is laid in contact with the original, and the two are heated by a lamp. The dark areas absorb more heat than the light areas, and a black image is formed directly on the copy paper.

17.11.2 Printing Processes

(1) Carbon Copies

When only a few copies—say half a dozen—of a report or letter are to be made for internal distribution, ordinary carbon-paper copies of the text may be adequate. Although carbon copies tend to smudge, they can be well produced, especially with some of the newer materials. Carbon copies are cheap and convenient when reproducing machines are not readily available.

(2) Spirit Duplicator (Ditto)

The spirit (or Ditto) process is convenient, because text and rough sketches can readily be put on the same page. The usual color is purple, but several other colors are available, and they can be mixed on the same page. The equipment is comparatively inexpensive.

On the other hand, the printed image is rather weak and fuzzy, and it tends to fade unless a special fluid is used. Not more than about 300 copies can be made from each master.

In the spirit process, a piece of special carbon paper that contains the pigment is put face up behind a paper master while the copy is typed or drawn on it. Pressure on the face of the master picks up a mirror image on the back. The image can also be transferred to the master from original copy by Thermofax. In the printing process this negative image is partially dissolved by an alcohol solution and transferred directly as a positive image to the copies.

(3) Mimeograph

Mimeograph presses are relatively inexpensive and easy to use. The printed image is black but somewhat fuzzy. Several thousand copies can be made from one stencil. It is very difficult to produce successful sketches on Mimeograph, and the paper that must be used is rather porous and weak.

In the Mimeograph process the copy is put onto a stencil that consists of a thin, porous sheet covered with a special wax that is impervious to ink. Typing on the stencil (with the ribbon inoperative) or scratching on it with a special stylus removes the wax from the areas where the image is to appear. Copy can also be imprinted on the stencil by an electrostatic or a thermal process. The stencil is then mounted on a drum that has an inked surface and rolled across the paper. Wherever the wax has been removed from the stencil, ink seeps through and prints on the paper.

(4) *Direct-image Offset*

The highest-quality large-scale office duplicating process is a form of offset lithography. The best-known direct-image-offset equipment is tradenamed Multilith.

Direct-image-offset printing is black and clear-cut; it is virtually as good as the copy from which it is made. Detailed drawings can be put right in with the text, and even halftone illustrations can be reproduced. There is no reasonable limit to the number of copies that can be made from one master. Any kind of paper and any color of ink may be used.

In this process the image (positive) is typed with a special ribbon or drawn onto an ink-repellent paper or plastic "plate." When an inked roller on the press goes over this plate, ink is deposited on the image and is repelled by the rest of the plate. The positive ink image is then picked up by another roller, on whose surface it presents a negative image. This roller is rolled over the paper, laying down a positive image. (The term *offset* refers to the intermediate process of picking up the image on a roller before it is impressed on the paper. If this step were omitted, and the original positive ink image were transmitted directly to the paper, it would appear as a negative.)

(5) *Photo-offset*

Photo-offset printing is not an office duplicating process, but a full-fledged printing-plant process. The printing operation itself is essentially the same as in direct-image-offset, except that the presses are much larger; but the image is put on the plate photographically instead of directly.

Photo-offset can produce very fine results; in fact it is used for printing many fine books and magazines. It is particularly flexible and convenient for combinations of text and line drawings, and it reproduces halftones very well.

Copy that has been prepared by typing, drawing, letterpress printing (see below)—or any other method—is photographed, with any desired amount of reduction or enlargement. The resultant negative is printed onto a zinc plate covered with a photosensitive emulsion. The surface of the plate is ink-repellent except for the area of the positive image. Copies are printed from this plate in an offset press just as in the direct-image-offset process described above.

(6) *Letterpress*

Letterpress printing can produce very fine results. It is not so convenient as photo-offset for mixtures of text and line drawings, and its plates usually cost considerably more than photo-offset plates. It is not economical unless at least several hundred copies are being run. Letterpress printing is rapidly being replaced by photo-offset.

In letterpress printing (also a printing-plant process), the ink is transferred to the paper from metal slugs or plates that have a negative image of the copy raised above the surrounding level. Text is set in movable type, usually by machine; and drawings or halftones are printed from acid-etched metal plates.

18. Tables

18.1 The Use of Tables

Numerical data presented in exposition form are very difficult to assimilate and interpret. For example, how much does this passage (adapted from an actual report) convey to you on first reading?

> Out of ninety shafts tested, sixty A7A's showed rusting, ten E7A's showed rusting, and ten each A7A and E7A did not.

Now take a look, even a quick one, at the same information tabulated:

Table 1. Rusting of Shafts

	Rusting	
Shaft Type	*Number With*	*Number Without*
A7A	60	10
E7A	10	10

At a glance, you take in all the numerical information and get a good idea of its significance. And much as this very small and simple set of data is clarified by tabulation, larger sets benefit even more.

When the really interested reader comes upon a prose presentation of data that he cannot readily understand, what does he do? He stops and tabulates the figures, either mentally or actually on paper, for himself; or perhaps he even plots them (see Chapter 19). It is more considerate and more efficient and more accurate for the author to supply the tabulation (or the plot) once and for all, for all of his readers. We urge you to use tables freely as a means of communicating information efficiently.

18.2 The Place of Tables

For the purposes of this book, we can divide tables into two main categories: those that form an integral part of the presentation of ideas, and those that are included simply for record purposes. The chief differences in the treatment of these two classes are that (1) the "record" table should usually be tucked out of the way in an appendix, while the "integral" table is most useful and most convenient if it appears at the appropriate place in the text; and (2) the record table may contain more figures than the integral table. We are most interested in the integral tables, but the principles discussed in this chapter apply to both kinds.

Tables are serially numbered, often in Roman numerals, and usually in a series separate from the series of Arabic numerals used for all the illustrative material, or *figures*. Sometimes, however, the tables are numbered in the main series of figures, and are labeled "Figure 10" rather than "Table X."

18.3 General Principles

18.3.1 Conciseness

Tables, like other forms of communication, are clearer and easier to understand if they are short. If a table you are constructing seems forbiddingly long, try breaking it up into two or more shorter ones. Or, if you must include a long table for record purposes, put it into an appendix, and quote brief excerpts from it in your text (see Secs. 7.2 and 7.4).

By all means, include no column of information in a table unless it has a direct bearing on your subject. It is almost safe to say, include no column of information unless it is specifically referred to in the text.

18.3.2 Titles and Headings

Every table should have a title. Like all other titles, it should be clear and as brief as it can be, yet still descriptive of the contents of the table (see Sec. 2.14). Although the reader may not get a *complete* understanding of the nature of the table without reading the accompanying text, he should get a general or basic idea of it from the title alone.

The conventional position for the title is above the table, with the table number immediately preceding the descriptive title. If the tables are numbered in the main series of figures, however, the title and figure number should go beneath the table, in the conventional position for figure titles.

Similarly, every column and every row in the table should have a clear, concise, descriptive heading. In order to save space, the words in a column heading may often be "stacked up," like those over the first two columns of Table 2.

18.3.3 Units

Never fail to name units of measurement. Usually the column or row heading is the convenient and efficient place to specify units; see Tables 2, 3, 7, and 10. On the other hand, if all the figures in the table are in the same units, the designation may be implicit in the main title. Sometimes you will see the name of the unit following each of the figures, as in Table 4. This position is likely to result in needless repetition, and is generally less workmanlike than the other two. However, in tables of sums of money, the dollar sign may well go right in the columns, preceding the figures in the first and last lines.

18.4 Some Specific Suggestions

18.4.1 Underlining of Column Headings

The most precise way of connecting column headings with the figures to which they apply is by means of vertical ruling, or boxing, as in Tables 4, 5, 6, and 7. But this boxing entails considerable complication and labor, particularly when the table is typewritten. An acceptably precise arrangement of column headings illustrated in Table 2 uses only horizontal ruling that can be easily made on the typewriter.

Table 2. Gasoline Analyses

Sample No.	Gravity, °API	ASTM Distillation, °F					Octane Number	
		IBP	10%	50%	90%	EP	Motor	Research
1	60.2	98	144	225	319	386	89	97
2	64.9	102	133	206	302	378	95	102
3	68.1	98	126	220	340	378	91	100
4	63.5	99	141	211	289	356	90	95
5	62.0	96	127	195	298	368	90	99

Notice in particular that the length of each of the horizontal rules shows the extent, or the number of columns, to which its heading applies.

A frequent practice is to run the horizontal ruling under headings all the way across the table, like this:

		ASTM Distillation, °F					Octane Number	
Sample No.	Gravity, °API	IBP	10%	50%	90%	EP	Motor	Research
1	60.2	98	144	225	319	386	89	97

Etc.

What each heading covers is now by no means so clear. The horizontal ruling should be put to some use besides simply separation of the headings from the body of the table.

18.4.2 Column Alignment

In a column of figures of like units, the decimal points should be lined up in the conventional way. But in a column of unlike units, particularly when the magnitudes differ greatly, it is wasteful of space and illogical to align the decimal points; the numerals should be centered in the columns, as shown in Table 3.

Table 3. Analysis of Used Oil

Sulfur, percent	0.22
Iron, percent	0.06
Lead, percent	0.006
Viscosity @ 100 F, Saybolt seconds	642
Viscosity @ 210 F, Saybolt seconds	73
Viscosity Index	104

18.4.3 Duplicate Entries in Adjacent Columns

It is best to avoid duplicate entries in adjacent parallel columns. You will often see tables constructed as in Table 4.

Notice that the differences between Car A and Car B fade into the background of duplicate entries. The distinguishing characteristics are by no means immediately apparent; you have to search them out.

Now look at this table rearranged (Table 5) with duplicate entries centered between the columns to which they apply.

A comparison of the two cars is now easy and automatic; it becomes immediately obvious that they are identical except for their engines and transmissions. The increased prominence of the differing entries suggests further revisions whose need was not nearly so apparent in the original table: the *weight* line might well be put beneath the *transmission speed* line, and the *tire mileage* line might be moved to the end of the table.

Table 4. Specifications of Rumbler Experimental Car

	Car A	*Car B*
Wheelbase	85 in.	85 in.
Tread	55 in.	55 in.
Overall length	160 in.	160 in.
Overall width	63 in.	63 in.
Weight	1650 lb	1750 lb
Seat width	50 in.	50 in.
Overall height	50 in.	50 in.
Head room	29 in.	29 in.
Leg room	28 in.	28 in.
Engine hp	60	100
Trans. speeds for'd.	3	4
Top speed	80–85 mph	90–95 mph
Fuel economy	30–35 mpg	25–30 mpg
Tire mileage	50,000 or more	50,000 or more
Approx. price	$2000	$2200

Entries duplicated in adjacent columns can mask not only *differences,* but also *similarities.* Take a look at Table 6 on page 228.

Now look at Table 7, a revised version of Table 6.

A lot of information about these engines that was buried in the original table makes itself almost immediately apparent in this revised version: There are two groups of engines, 6-cylinder and 8-cylinder. All the 6's apparently use the same block (same bore), but crankshafts of two different strokes provide engines of two displacements. Model 160 is apparently the same as Model 150 except that its higher compression ratio produces greater power output. Each of two 8-cylinder blocks comes with two different strokes, providing engines of four displacements and four power ratings.

Caution: The device of centering entries between columns does not work clearly unless the table is completely boxed.

Table 5. Specifications of Rumbler Experimental Car (Revised)

	Car A	*Car B*
Wheelbase, in.	85	
Tread, in.	55	
Overall length, in.	160	
Overall width, in.	63	
Weight, lb	1650	1750
Seat width, in.	50	
Overall height, in.	50	
Head room, in.	29	
Leg room, in.	28	
Engine hp	60	100
Trans. speeds for'd.	3	4
Top speed, mph	80–85	90–95
Fuel economy, mpg	30–35	25–30
Tire mileage	50,000 or more	
Approx. price	$2000	$2200

Table 6. Rumbler Truck Engines

Model	*140*	*150*	*160*	*170*	*180*	*190*	*200*
Cylinders	6	6	6	8	8	8	8
Bore, in.	3⅞	3⅞	3⅞	3⅞	3⅞	4	4
Stroke, in.	3½	4⅛	4⅛	3¼	3½	3½	3¾
Displacement, cu in.	250	292	292	307	330	350	400
Compression ratio	8.5	8.0	8.8	9.0	9.0	9.0	9.0
Max. gross hp	155	170	180	200	225	255	310

Table 7. Rumbler Truck Engines (Revised)

<table>
<tr><th>Model</th><th>140</th><th>150</th><th>160</th><th>170</th><th>180</th><th>190</th><th>200</th></tr>
<tr><td>Cylinders</td><td colspan="3">6</td><td colspan="4">8</td></tr>
<tr><td>Bore, in.</td><td colspan="3">3⅞</td><td colspan="2">3⅞</td><td colspan="2">4</td></tr>
<tr><td>Stroke, in.</td><td>3½</td><td colspan="2">4⅛</td><td>3¼</td><td colspan="2">3½</td><td>3¾</td></tr>
<tr><td>Displacement, cu in.</td><td>250</td><td colspan="2">292</td><td>307</td><td>330</td><td>350</td><td>400</td></tr>
<tr><td>Compression ratio</td><td>8.5</td><td>8.0</td><td>8.8</td><td colspan="4">9.0</td></tr>
<tr><td>Max. gross hp</td><td>155</td><td>170</td><td>180</td><td>200</td><td>225</td><td>255</td><td>310</td></tr>
</table>

18.4.4 Duplicate Entries in Successive Lines

You can usually replace columns of ditto marks (or literally repeated entries) by a more informative scheme. Table 8 is adapted from a report on some experimental joints.

Table 8. Performance of Experimental Joints

Joint No.	Position	Type Weld	Hours to Failure
90	1L	Butt	1540
91	2R	"	1734
92	3L	"	1307
93	4R	"	1915
94	5L	"	2060
95	1R	Lap	2758
96	2L	"	2592
97	3R	"	2621
98	4L	"	3414
99	5R	"	2866

This arrangement masks the fact that one of the important purposes of the test was to compare the durability of the two types of welds. One way to clarify this table is to remove the *type-weld* column and substitute horizontal subheads for it, as in Table 9. The more obvious separation of the joints into two groups suggests the worth of the two *average-hours* entries, which have been added.

Note that the first horizontal subhead goes *below* the column headings. (For some reason that has no basis in logic, you will sometimes find the first subhead above the column headings.)

Another effective substitute for the column of dittos is illustrated in Table 10. This scheme requires boxing.

Table 9. Performance of Experimental Joints (Revised)

Wheel No.	Position	Hours to Failure
	Butt-welded Joints	
90	1L	1540
91	2R	1734
92	3L	1307
93	4R	1915
94	5L	2060
Avg		1711
	Lap-welded Joints	
95	1R	2758
96	2L	2592
97	3R	2621
98	4L	3414
99	5R	2866
Avg		2850

Table 10. Performance of Experimental Joints (Revised)

Type Weld	*Joint No.*	*Position*	*Hours to Failure*
Butt	90	1L	1540
	91	2R	1734
	92	3L	1307
	93	4R	1915
	94	5L	2060
	Avg		1711
Lap	95	1R	2758
	96	2L	2592
	97	3R	2621
	98	4L	3414
	99	5R	2866
	Avg		2850

18.4.5 Adjacent Columns for Numbers to Be Compared

In a table containing sets of numbers that are to be compared, you can ease the reader's task, and make the comparison more evident, by arranging them in adjacent columns rather than stringing them out in a single column. Table 11, from a report on possible schemes for a manufacturing process, obscures comparison.

The same data rearranged in adjacent columns, Table 12, permits an easy comparison of the costs of the various schemes.

Table 11. Cost Per Record

Single-opening Press	
Coating Contracted	
1 shift	$0.0032
2 shifts	0.0030
Coating in Shop	
1 shift	0.0029
2 shifts	0.0027
Double-opeing Press	
Coating Contracted	
1 shift	0.0031
2 shifts	0.0029
Coating in Shop	
1 shift	0.0028
2 shifts	0.0026

Table 12. Cost Per Record (Revised)

	Coating Contracted		Coating in Shop	
	1 Shift	2 Shifts	1 Shift	2 Shifts
Single-opening press	$0.0032	$0.0030	$0.0029	$0.0027
Double-opening press	0.0031	0.0029	0.0028	0.0026

19. Visual Presentation of Information

19.1 The Use of Visual Presentation

Very important tools for technical reporting are the visual or graphical aids—pictures, diagrams, and graphs. Often a glance at a schematic diagram will reveal an arrangement, or a glance at a curve will show a relationship, that words alone could never make really clear. You should by all means use the visual aids freely whenever they will help your reader.

The classical view was that graphs and drawings were strictly to *supplement* the verbal presentation of information. But if you can give your reader a grasp of a complicated machine most clearly and most easily by means of a drawing, why burden him with a lengthy word description besides? If a curve that shows the relationship between some data also shows their numerical values with sufficient accuracy, why clutter up your report with a table of those same values? A growing tendency is to let visual presentations *supplant* verbal ones as well as supplement them.

But if your verbal exposition is to be logically complete, it must contain at least the fundamental description, or definition, discussed in Sec. 2.4. For instance, it is insufficient to say simply:

> The essential characteristics of the d'Arsonval movement are shown in Fig. x.

A statement that is not very much longer but a great deal more satisfying to the reader might run this way:

> The d'Arsonval movement, as shown in Fig. x, is basically a coil pivoted in a permanent magnetic field. A torque proportional to current flowing in the coil is resisted by springs. A pointer attached to the coil indicates the displacement of the coil and thus the magnitude of the current.

Fig. x, of course, fills in a lot of details not covered in this basic definition.

19.2 The Place of Visual Presentation

In most technical reports and papers the figures are an integral part of the presentation. Therefore the most useful place to put them is right with the text they illustrate. But if they are inconveniently large or bulky, or if one or more of them is referred to several times in different parts of the text, it may be more convenient to group them all together following the text. If you refer to a figure that is neither nearby nor in a group at the end, specify not only the figure number but also the page number. Figures, such as working drawings, that are included only for record purposes can well be put at the end, in an appendix.

All the figures—graphs, drawings, diagrams, photographs, maps—are usually numbered in one series of Arabic numerals. Sometimes in a long work such as a textbook a separate series is used for each chapter, with the chapter number preceding each individual number, thus: 1-1, 1-2, 1-3; 2-1, 2-2, 2-3; etc. Whatever system you use, be sure that you never assign the same number to two different figures in one report or paper.

19.3 Reference to Figures

Remember that every figure must be integrated into the verbal presentation by means of a specific reference in the text. The position of this reference is very important:

(1) *The reference should precede the figure whenever possible.* A drawing or curve that has not been mentioned in the text interrupts the reader's train of thought; it mystifies and annoys him. He is likely to overlook its significant features. On the other hand, if the figure comes after the reference the reader will obey instructions to refer to it at the appropriate time. He will not be interrupted, and he will know what to look for in the figure.

(2) *The reference should be made as early in the discussion as it appropriately can be.* To learn at the end of a complicated and tedious description that "the details of this construction are clearly shown in the drawing on the next page" is of very little help, because it comes too late. The right time for the reader to look at the drawing is while he is reading the description.

19.4 General Principles

19.4.1 Simplicity

Although their purpose is usually to clarify and facilitate the communication of information, the visual aids can suffer from the same fault that characterizes so many technical reports: the burial of the main, basic idea under a lot of details. When you construct a graph or a drawing, do not obscure the main idea by surrounding it with unnecessary details. *Make your graphical presentations as simple and as open and as bold as you can.*

Sometimes the size of the figures that you prepare (or obtain) will be reduced in the published version of your paper or report. There is always the danger that lettering or other details will not be legible after the reduction. Therefore, when you are preparing figures for publication, try to plan them specifically for reduction to the final size. Be sure that the lettering is large enough to be easily legible after reduction. To make all the drawings in a paper or report look uniform, standardize on one reduction factor and one series of letter sizes for all the drawings.

19.4.2 Titles; Independence

Every figure should have a descriptive title in addition to the figure number. Number and title are usually placed beneath the figure. How much information should the title supply? Certainly it should describe the basic nature of the figure, if for no other reason than for the benefit of those "readers" who look only at the pictures. But how much more information should it carry?

Some organizations insist that every drawing and every graph or curve sheet be able to "stand alone"; that is, be *entirely comprehensible* to someone who has only that sheet, and no accompanying text or explanation. If this criterion is to be met, each sheet must supply a lot of background, or orientation, information in addition to its specific title. For instance, if it were being sent out by itself, the curve sheet of Fig. 6 on page 242 would carry some or all of the following information in addition to what it has: car numbers; weight and axle ratios of cars; odometer readings; dates of tests; temperature, wind direction and velocity, and weather conditions during tests; grade of oil in engines; and fuel used. If the sheet is going to be distributed separately—by itself—then it needs this sort of information in order to be meaningful. On the other hand, if the graph or drawing is going to be bound into a report (or published in a paper) that supplies the background information anyway, why clutter up the page with nonessential material?

Every figure title should be no longer than necessary to make the figure readily comprehensible *in the medium where it is published*. Thus, if you are going to use in your report or paper figures that were originally prepared to stand alone, you will do well to remove most of the title information that also appears in nearby text. Graphs that are to be published in important reports or papers will probably have to be redrawn anyway, for reasons discussed in Secs. 19.5.6 and 19.5.7.

19.5 Graphs and Curves

19.5.1 The Quick Look

Just as tables communicate some kinds of ideas more rapidly and more easily than exposition does, graphs and curves are often more effective and more efficient than either exposition or tables. The special virtue of graphs is that they present information very rapidly—at a glance—so that the viewer can grasp a number of values almost simultaneously. As a result, he is readily able to compare these values and to get an idea of their *relationships* with each other.

Consider this simple table of automobile fuel economy obtained from a proving-ground test:

Table 13. Constant-speed Fuel Economy on Level Road
1970 Rumbler

Speed, Miles per Hour	Fuel Economy, Miles per Gallon	
	6-Cylinder Engine	8-Cylinder Engine
20	25.1	23.0
25	25.3	23.0
30	24.8	22.8
35	24.0	22.6
40	23.2	22.0
45	22.0	21.2
50	20.9	20.0
55	19.3	18.9
60	17.3	17.2
65	14.8	15.2
70	12.0	13.0
75	9.4	10.2
80	6.5	7.9

You can see that both engines consume more fuel as car speed increases; but at what speeds does this tendency become pronounced, and how do the two engines compare? That is, "What is the shape of the curve?"

Now look at Fig. 8 on page 244. You see immediately that (1) the falling off in fuel economy becomes rapidly more pronounced at higher speeds; and (2) the 6-cylinder engine is more economical than the 8 at low speeds but less economical from about 60 miles per hour up.

You can read the values of the fuel economy curves of Fig. 8 to perhaps ½ mile per gallon. The scatter of the observed points indicates that closer readings are not warranted; and besides, smaller differences are not really significant. Therefore, why not present the results of the fuel-economy test only in curve form, without any tabulation?

All of the graphical forms for presenting data show relationships; some of them stress the relationships; others simply list facts. For our purposes, graphs can be divided into three main categories:

(1) *The curve* is most useful in showing relationships among continuously varying phenomena; at the same time it presents facts.

Engineers and scientists use the curve more than any other kind of graph. Although the principles discussed in the rest of Sec. 19.5 pertain directly to curves, most of them may be applied also to bar charts.

(2) *The bar chart* presents information about discontinuous phenomena or discrete quantities—the sort of information that is often simply listed in tables. The bar chart makes it particularly easy to *compare* data. It is a valuable tool that engineers could profitably use oftener than they do.

A special adaptation of the bar chart is the pictorial chart, in which quantities are represented not by the lengths of bars, but by the lengths of rows (and thus by the number) of little pictorial symbols, each of which stands for a stated number of units. The pictorial chart is effective in presenting numerical data to laymen, who are not generally conversant with graphical representations. But scientists and engineers will not be baffled by bar-chart representation, and they may be annoyed by the naïveté and lack of precision of the pictorial chart. Therefore, when your audience is technically trained, you will probably do better to use ordinary bar charts rather than pictorial charts.

(3) *The pie chart* is an effective and highly specialized device for showing just one thing—proportion, or percentage, of the whole. Although it is used in business reports more than in technical reports, the pie chart should not be overlooked by technical men.

19.5.2 Direction of Variables

According to a well-established convention, the independent variable is plotted horizontally, the dependent variable vertically (except that bar charts are sometimes constructed with the bars running horizontally). You should follow this convention if only because your readers will be momentarily confused if you do not. But besides, it is a convention that has a good, logical basis. We are primarily interested in values of the dependent variable, and it seems natural to express magnitude by height.

19.5.3 Scale Labels

Remember to furnish complete scale information along the sides or axes of every graph. Unless a curve is used to show only the rough shape of a relationship, each scale label contains two pieces of information: the name of the variable and the units in which it is measured.

These two pieces of information are best separated by a comma, thus:

Frequency, megacycles

Another acceptable method is to put the units in parentheses, thus:

Height above sea level (feet)

The word *in* will serve as a separator, but it takes up additional space:

Current in amperes

Some people separate the name of the variable from the units by a dash:

Change of free energy—calories

But because the dash may occasionally be taken for a minus sign, you should get in the habit of avoiding this notation.

You have probably seen scales labeled with multipliers, like this:

Force, lb × 1000
or:
Force, lb ÷ 1000

Such multipliers do save space by eliminating a lot of zeros from the numbers along the axes, but they result in at least momentary ambiguity. The reader wonders, "Have these numbers *been* multiplied (or divided) by a thousand, or am I *supposed to* multiply them (or divide them) by a thousand?" His knowledge of the subject under discussion will probably enable him to answer the question after a little thought, but you should not put him to this trouble. Better, express your multiplier in an unambiguous way:

Force, thousands of lb

19.5.4 Choice of Scale

(1) *Rational Scale*

Always choose a rational scale—that is, one that neither underemphasizes nor overemphasizes the importance of the changes you are depicting. For instance, the curve of Fig. 2 gives the impression that the average compression ratio of U.S. automobile engines dropped drastically from 1959 to 1961, then increased rapidly but sporadically from 1961 to 1968. The same data plotted to different scales in Fig. 3 seem to indicate that there was a very slight rising trend from 1959 to 1968 with a small reversal at the beginning of the period. A truer representation would be given by scales somewhere between the ones used in these figures.

There is no formula for the choice of rational scales. You simply have to look at the slope of your curve and consider whether it seems to give an accurate impression of the idea you are trying to convey.

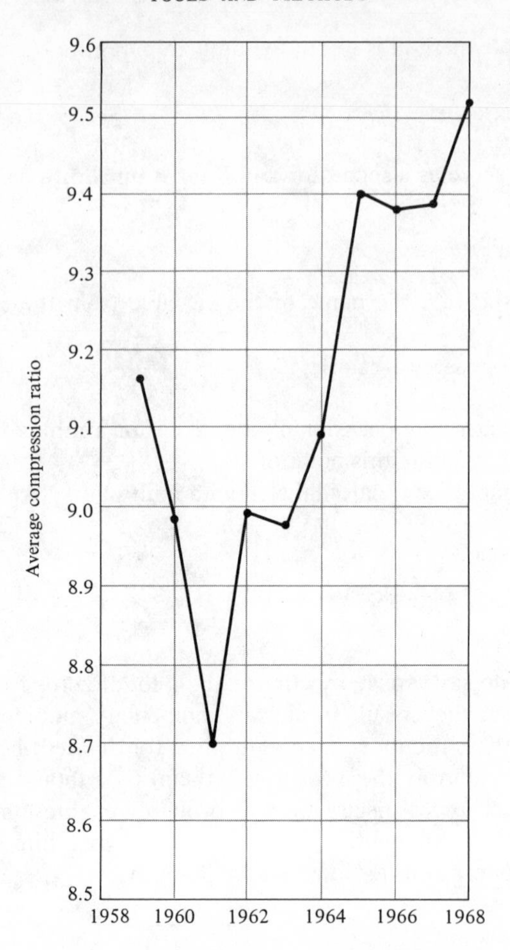

Fig. 2. Trend in average compression ratio (changes overemphasized)

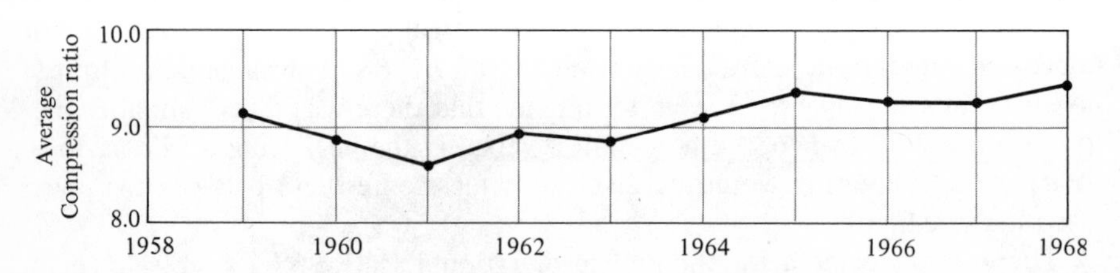

Fig. 3. Trend in average compression ratio (changes underemphasized)

(2) *Values for Scale*

For ordinary decimal graph paper (each main division divided into 10 subdivisions) the value of each main division should be restricted to 1, 2, or 5 and their multiples by any power of 10. Thus you can let each main division represent 0.01 or 1 or 100 or 1,000,000; 0.2 or 2 or 200; 0.005 or 5 or 5000; and so on.

Points plotted according to this rather arbitrary-sounding rule can be read almost by inspection. For instance, see how quickly you can read the points in Fig. 4.

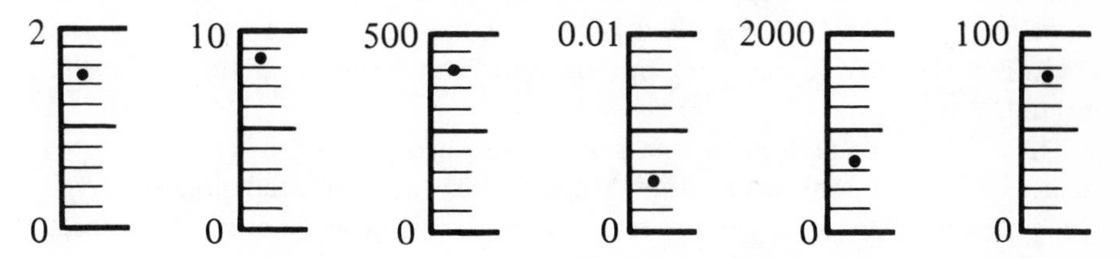

Fig. 4. Points plotted to permissible scales

But in a misguided effort to make the curve look pretty, or to fill up the paper, some people assign other values to the main divisions of decimal graph paper. Now try to read the points of Fig. 5 rapidly.

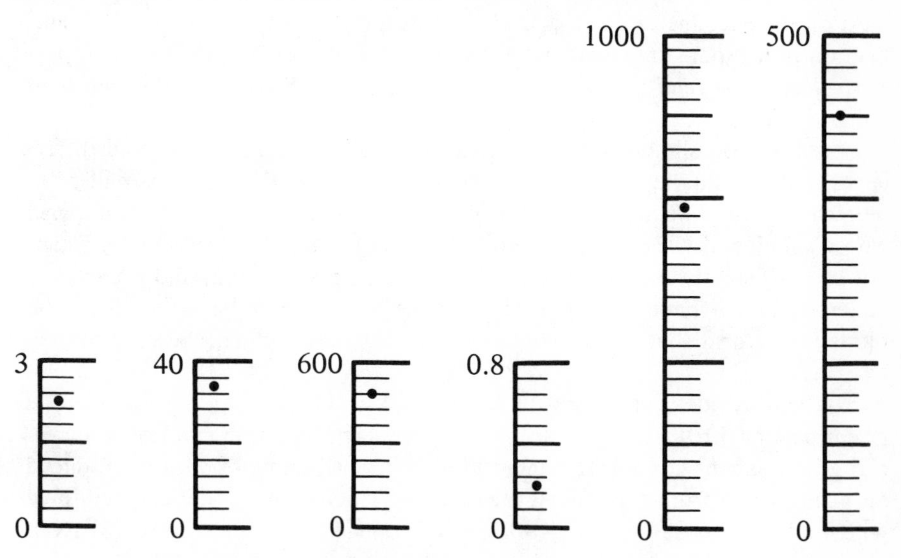

Fig. 5. Points plotted to prohibited scales

And remember that these scales will not only bother your reader, but will also slow you down while you are doing the plotting. Better stick to the "permissible" scales.

19.5.5 Bold Curves

When you plot a curve on commercial graph paper—which is usually printed in a shade of red or green—the curve is clearly distinguished from the lines of the grid by color separation. But when this sheet is reproduced in one color by Ozalid or any other photographic process, color difference no longer separates the curve from the grid, and they tend to become confused, as shown in Fig. 6, page 242.

If you plan to make one-color prints from your graphs (as is done for practically all technical reports), *you should make the curves at least 3 times as thick as the thickest grid lines.* In Fig. 7 the curves of Fig. 6 have been made considerably thicker; notice how they stand out from the grid background.

There is one exception to this rule: if you are plotting a *working* curve, such as a calibration curve, from which values must be read precisely, you will probably have to use a fine line and hope that it will never be confused with the grid lines.

19.5.6 Coarse Grid

The commercial graph paper most often used has 20 divisions to the inch. This fine grid is often confusing; it is usually unwarranted; and it is always hard on the eyes (see Figs. 6 and 7). Whenever you prepare a graph for publication in a report or paper, by all means *use the coarsest grid that will permit values to be read with the appropriate precision.* Sometimes you can omit the grid entirely.

Certainly you should not use a grid that makes it possible to read differences that are smaller than the experimental error or that are so small as to be meaningless. For instance, we have seen octane numbers plotted on a grid whose smallest division represented 0.05 number. Now, octane numbers cannot be consistently measured to less than 0.5 or perhaps 0.2, and differences of less than 0.5 have little real significance. Octane numbers should not be plotted on a grid that has divisions finer than 0.5 (which can be read to perhaps 0.2).

You can avoid the fine grid in several ways: (1) You can use commercial graph paper with 1, 2, or 10 divisions to the inch. (2) You can use commercial graph paper with a grid printed in blue ink, which will not be reproduced on prints made from this paper. Before the prints are run, you can reinforce with India ink those grid lines that are to appear on the final copy. (3) Even if you prefer to do your preliminary plotting on ordinary 20-to-the-inch graph paper, you can trace your final copy on plain tracing paper, inking in only the wanted grid lines. Fig. 8 is a traced copy of Fig. 6. Notice that it is very much easier on the eyes than the versions with fine grids; yet it permits the fuel economy to be read to ½ mile per gallon, the greatest precision warranted by the nature of the data.

19.5.7 No Lettering on Grid

Try to avoid having any lettering superimposed on the grid lines. In Fig. 6 the lettering, both of the scales and the internal labels, is typed right on the grid. Color separation made it reasonably legible on the original sheet. But on this one-color reproduction, confusion with the grid lines makes the lettering difficult to read.

In Fig. 7 the scales are placed in the clear margin outside of the grid area, and clear spaces have been provided within the grid area for the internal labels. The lettering is much more easily legible than in Fig. 6. (But see the caution about binding margin in Sec. 17.9.)

Clear areas for lettering may be provided in several ways, depending on the kind of paper and the method of reproduction to be used:

(1) If a commercial graph sheet is to be reproduced by photolithography or photography (but not by direct printing), you can do your lettering on strips of white paper and paste them over the grid.

(2) If a commercial graph sheet is to be reproduced by direct printing, you can do your plotting and lettering on the back of the paper, and then with a blade scrape away the grid from behind the lettering.

(3) If you are tracing your curve sheet on plain paper—no matter how it is going to be reproduced—you can simply omit the grid lines wherever there is lettering inside the grid area; see Fig. 8.

To comply with the principles of this section and the preceding one, you should probably trace on plain paper most curves that are going to be published in any but very informal reports.

19.5.8 Label for Each Curve

Whenever more than one curve is plotted on one set of coordinates, it must be possible (1) to distinguish the curves and (2) to identify them. As shown in Figs. 6 and 7, the curves are usually distinguished by the use of differing symbols and lines. They are often identified by a *key* like the one in these two figures.

With this device the identification process is a double one, involving excursions of the eyes back and forth between the key and the curves in order to connect the names with the codes. A more direct method for identifying the curves—individual labels—is illustrated in Fig. 8. Individual labels are much more helpful than a key. You should use them unless physical complications on the curve sheet prevent.

19.5.9 Suppressed Zero

When the scale of the independent variable starts at any value above zero, it is said to have a *suppressed zero*. Because the function of graphs is to present information at a glance, the suppressed zero can be very misleading. Look at the bar chart (Fig. 9, page 245) from the annual report of a corporation.

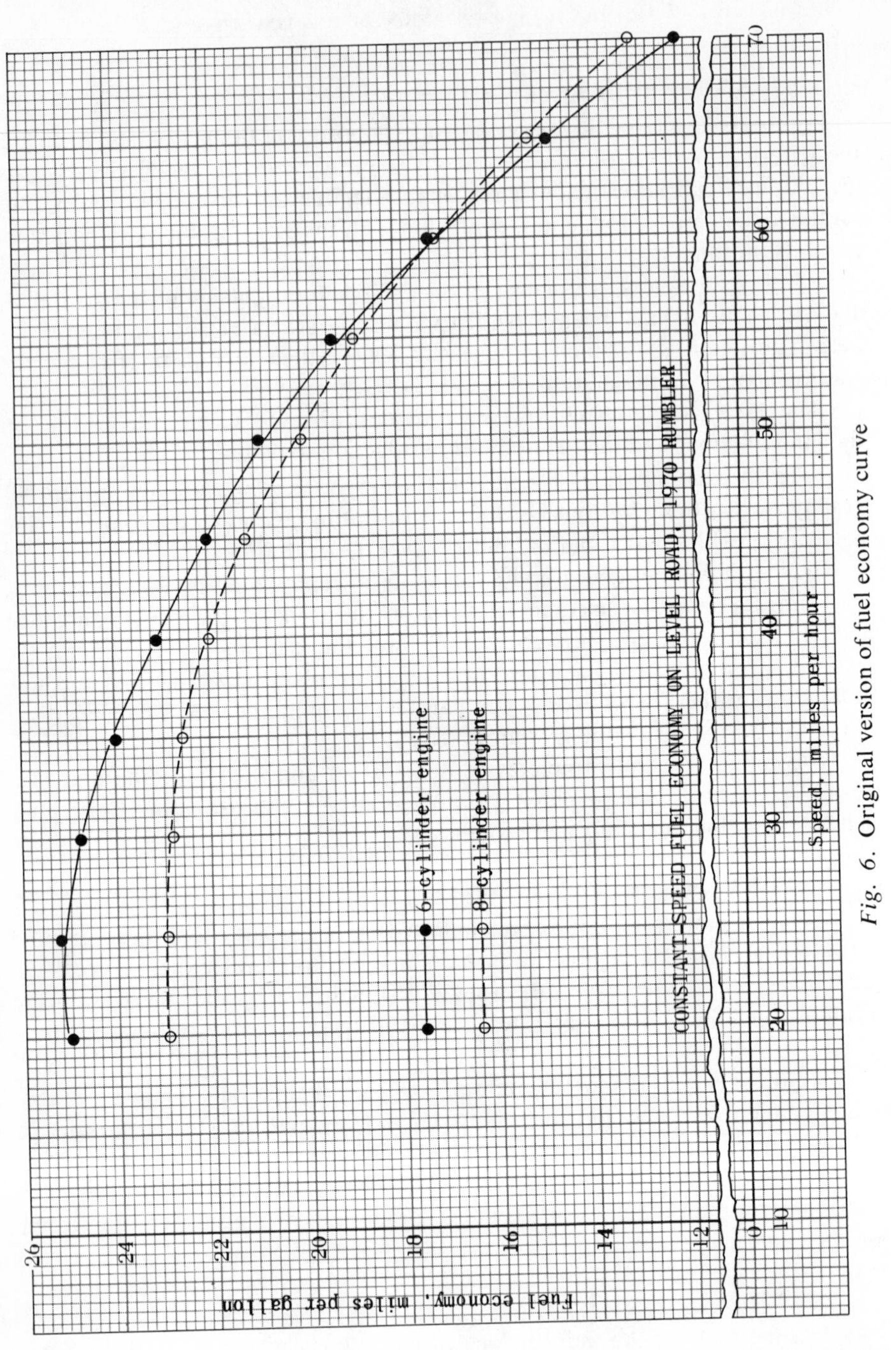

Fig. 6. Original version of fuel economy curve

NOTE: So that they would fit the pages of this book without being reduced or scaled down, Fig. 6 and its revised versions (Figs., 7, 8, 15, and 16) have been *condensed*. That is, part of the vertical scale has been removed (see Sec. 19.5.9), and the curves have been cut off at 70 miles per hour. In reality, all of these curves would be plotted on

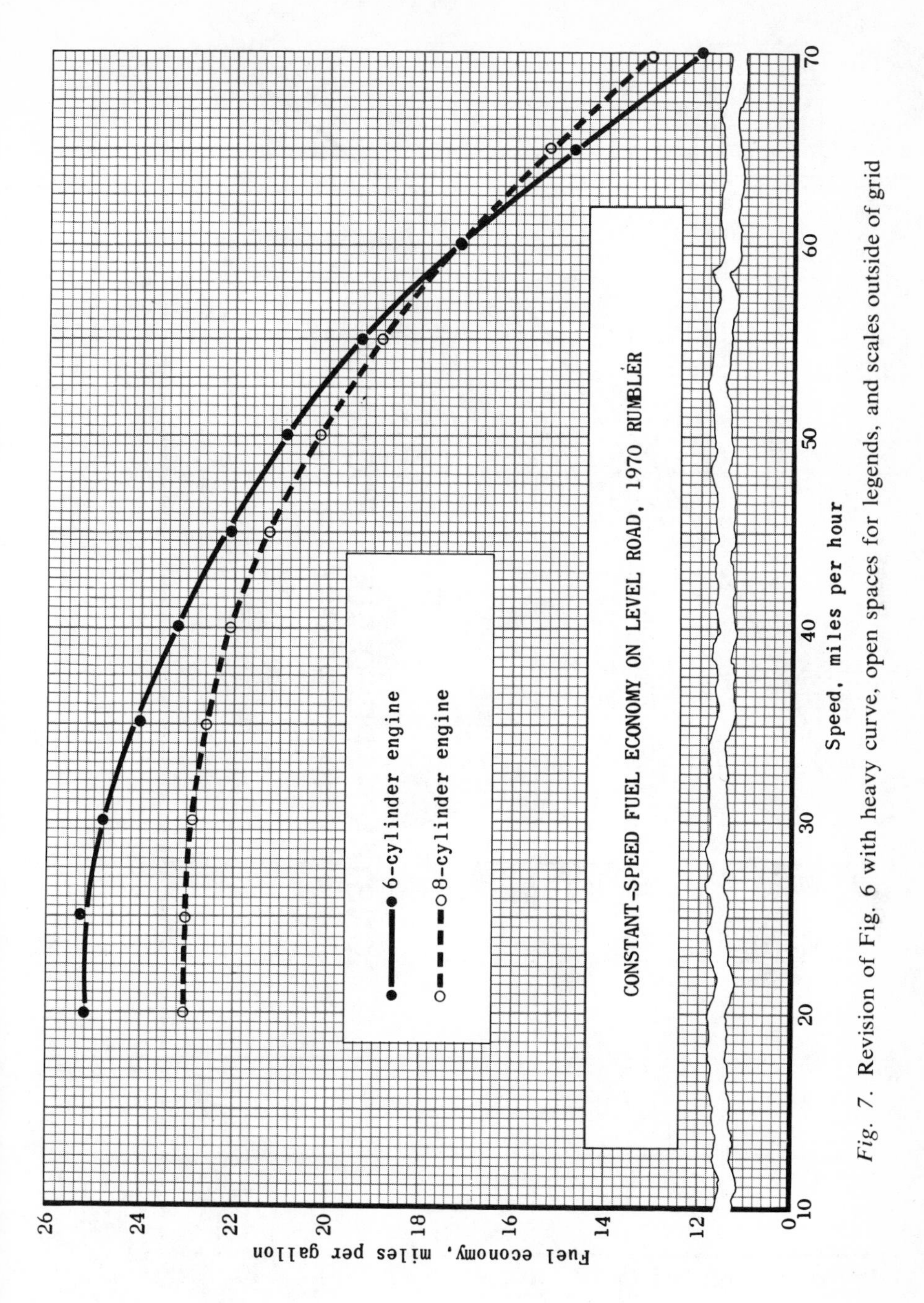

Fig. 7. Revision of Fig. 6 with heavy curve, open spaces for legends, and scales outside of grid

8½ x 11-inch paper; the vertical scales would run from 0 through 26 miles per gallon without a break; and the horizontal scales and the curves would be extended to 80 miles per hour.

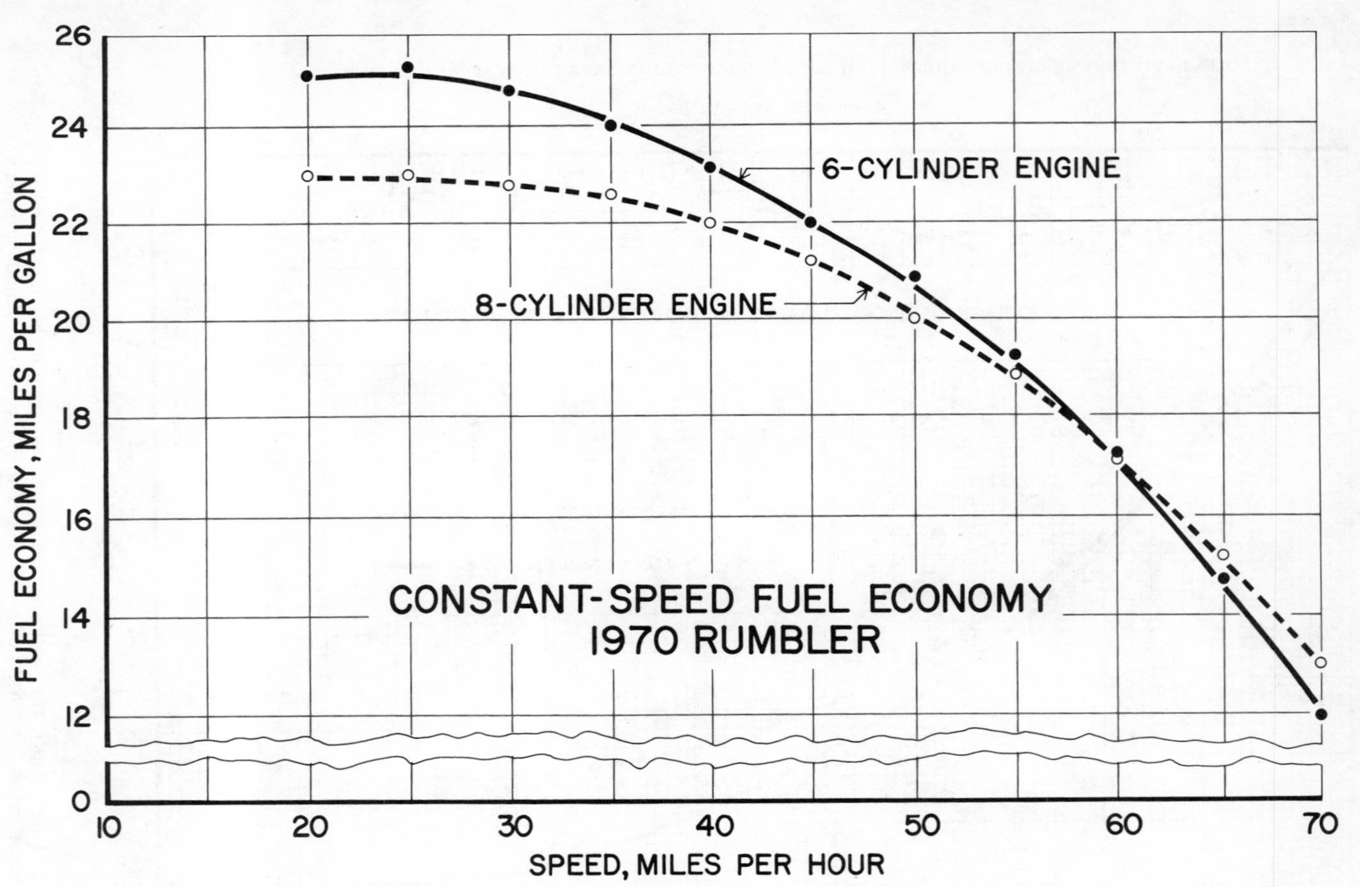

Fig. 8. Revision of Fig. 7 with only major grid lines traced, key replaced by labels

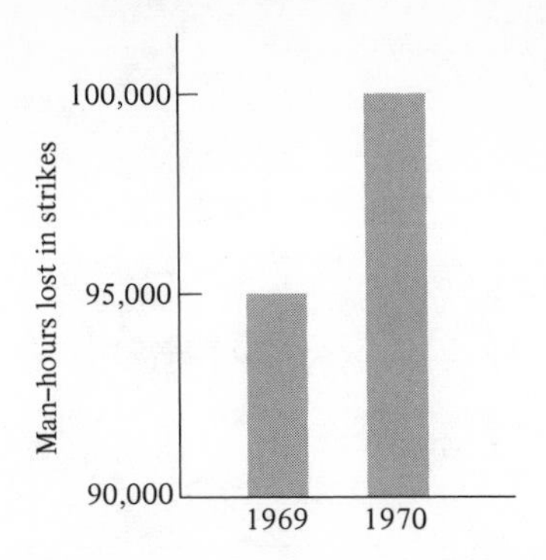

Fig. 9. Difference exaggerated by suppressed zero

The quick comparison so easy with a bar chart gives you the impression that the corporation lost twice as many hours in strikes in 1970 as it did in 1969. A more careful examination reveals that the 1970 loss was only about 5 percent greater than the 1969 loss.

The suppressed zero is sometimes used with the intention of misleading. This is an inherently dishonest practice.

But there is often a perfectly good justification for the suppressed zero. If the vertical scale of Fig. 9 had been brought down to zero, the graph would have looked like Fig. 10.

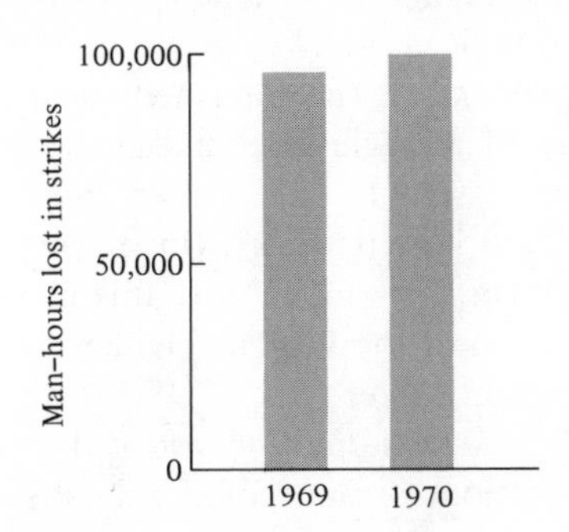

Fig. 10. Significant difference obscured

Now a significant difference—5000 man-hours—is hardly discernible. It is legitimate to suppress the zero so that a difference can be read; but whenever you do so, be sure to announce the fact plainly. Fig. 11 illustrates one way of indicating a suppressed zero.

It is not necessary to call attention to the suppressed zero when you are representing ratios or arbitrary units such as degrees of temperature or octane numbers, in which the zero has no real significance (see Figs. 2 and 3).

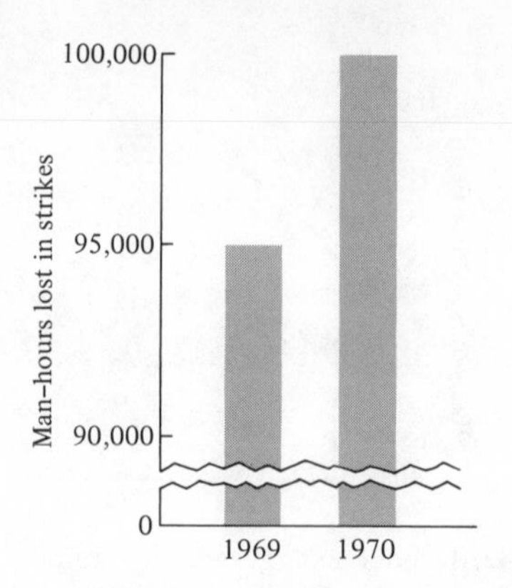

Fig. 11. Suppressed zero indicated

19.6 Other Forms of Visual Presentation

Information may be transmitted by numerous visual means besides graphs and curves. Drawings, diagrams, schematics, photographs, exploded views—even models—all make their contribution to the efficient communication of ideas.

These aids are usually produced by specialists, and a full discussion of them is outside the scope of this book. But we do want to enlarge upon the plea for simplicity voiced in Sec. 19.4.1.

Fig. 12 is an x-ray pictorial of the intermediate-frequency and audio-frequency amplifier sections of a single-sideband amateur receiver. If you are a radio technician, you can assemble the module from this pictorial, and you can undoubtedly understand how it works. But if you are not a radio technician or an electrical engineer, how much does this pictorial tell you about the essential nature and function of the various elements of the module and their relationships with each other?

Now consider Fig. 13, a *schematic diagram* of the same module. The simplification and the elimination of many of the details of the pictorial clarify the general scheme of the apparatus; but still, unless you are used to reading cicuit schematics you probably cannot tell how the module works.

Finally, look at Fig. 14, a *block diagram* of our module. Here the main elements are represented simply by blocks, and their *logical relationships* with each other are indicated by arrows. Notice how very quickly you can grasp the basic scheme of the module. A block diagram like this conveys intelligible information even to a layman; and to a technical man it describes fundamental arrangements more rapidly and more easily than any other means.

A consideration of these figures leads to an important general rule: *Always use the simplest—or least detailed—form of representation that still supplies those details needed for your immediate purposes.*

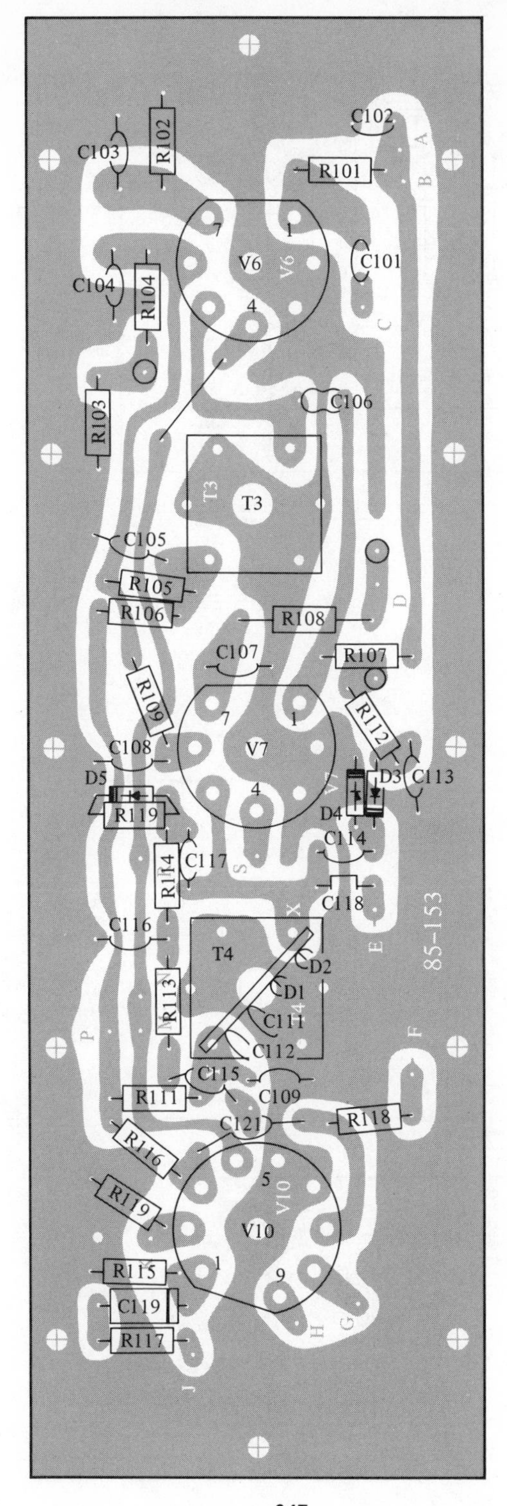

Fig. 12. An x-ray pictorial of radio receiver module

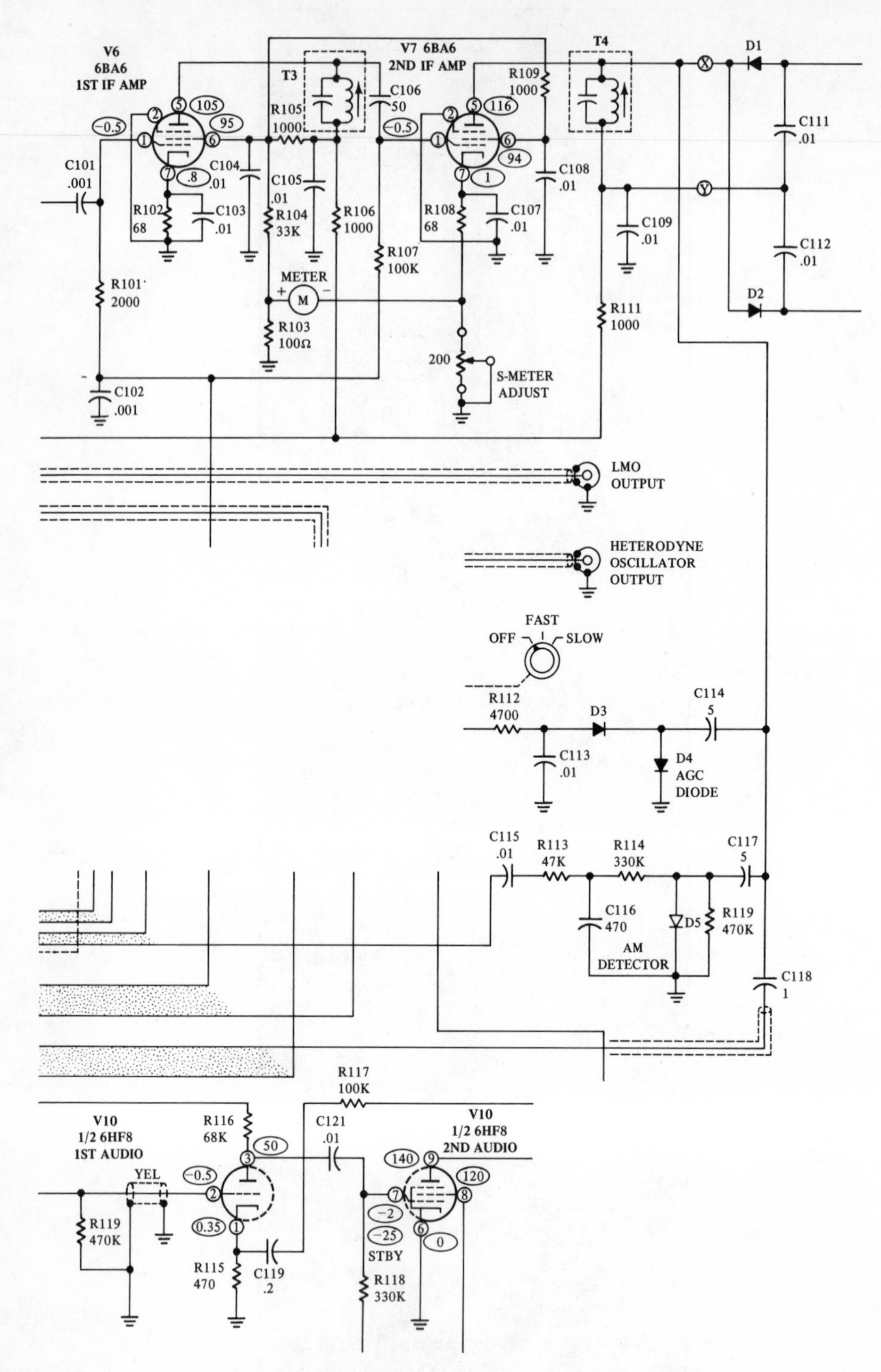

Fig. 13. Schematic diagram of radio receiver module

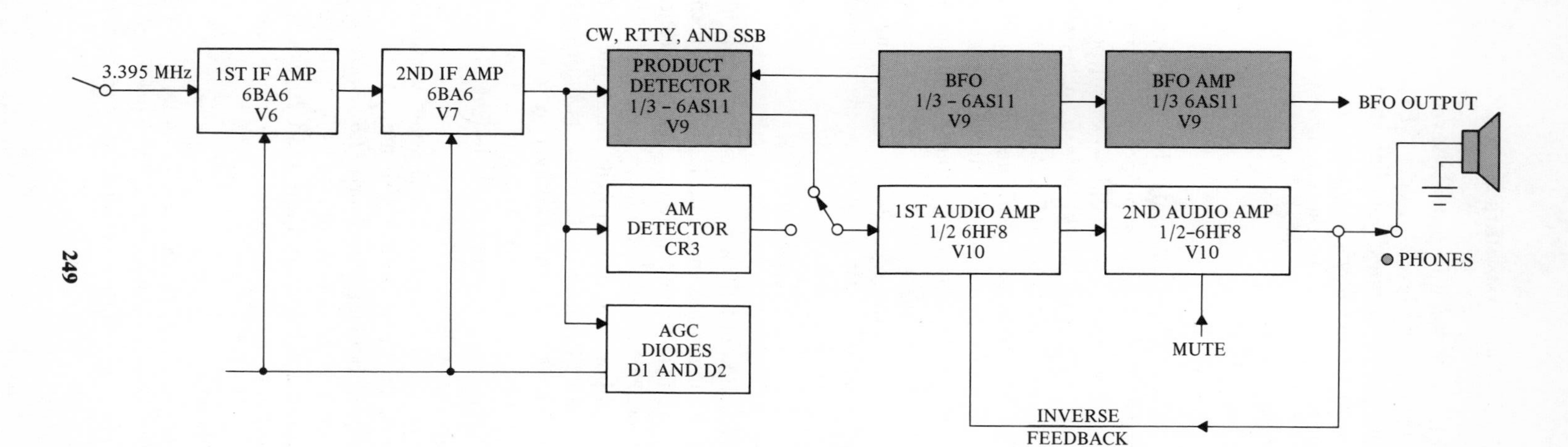

Fig. 14. Block diagram of radio receiver module

Quite often you will have occasion to provide several representations of an idea or an apparatus at increasing levels of detail. Even though you must finally supply complete working drawings, you may start out by presenting the basic, overall scheme in a simple block diagram or flow diagram. Between this and the final drawing may be one or more diagrams on intermediate levels.

19.7 Slides

Any of the visual aids, including tables, may be projected onto a screen to accompany the oral presentation of papers, proposals, or reports (see Chapter 13). Projection is most often by means of slides—small positive transparencies. Slides are still occasionally made in 3¼ × 4¼-inch size, but more often nowadays on standard 35-millimeter film in 2 x 2-inch mounts.

Visual aids may also be projected by means of an *overhead projector,* which uses 8½ x 11-inch transparencies made directly from positive copy; or an *opaque projector,* which projects directly from opaque positive copy such as the page of a book or report (see Sec. 13.1.2).

The principles that follow in the rest of Sec. 19.7 apply to all projected material. But since slides are the most usual form, the discussion is in terms of slides.

19.7.1 Simplicity

When a speaker uses a slide to help him communicate his ideas to an audience, he seldom stops talking while the slide is on the screen. If a slide requires extended study, the listener has the choice of comprehending the slide or listening to the speaker. He cannot do both.

We have stressed the importance of simplicity in all the visual aids; but in slides it is doubly important. Unless a slide presents information at a glance, it defeats its own purpose. A slide should have on it no detail that is not needed, no data that is not referred to in the talk. Every superfluous line should be removed; every line that remains should be clear, and the important lines such as curves should be extra bold.

These general ideas have been codified into some useful specific rules by the professional societies, at whose meetings so many slides are thrown upon the screen. In its helpful booklet *Handbook for Authors,* the American Chemical Society says that each slide should be limited to the presentation of one idea; that a slide should contain not more than 20 words, and preferably not more than 15; and that a table should have no more than 25 or 30 data.

19.7.2 The 1/40 Rule

In a bulletin on lantern slides, the Society of Automotive Engineers propounds the highly valuable rule that *no letter or figure shall be less than* $^{1}/_{40}$ *the vertical height of the copy.* Main titles and other important words can, of course, be proportionately larger. Application of this rule results in lettering that shocks most draftsmen on first acquaintance; but it has been a godsend to the back rows of technical-society audiences.

The 1/40 rule assumes that the height of the copy is its limiting dimension. If the ratio of width to height is greater than the 3/2.25 of the standard slide, then the width becomes the limiting dimension. In this event, the hypothetical height is 2.25/3 of the actual width, and the smallest characters should be at least 1/40 of this hypothetical height.

If you draw your slides horizontally on 8½ x 11-inch paper so as to fill up most of the sheet, the *smallest* characters should be made with a 175-size (0.175-inch) Leroy or Wrico guide. It is a good idea to standardize on 8½ x 11-inch paper for slide drawings. This practice will produce slides of uniform appearance; and any slide that cannot be conveniently drawn on a sheet of this size very probably has too much on it.

The empirical 1/40 rule leads to an important corollary: *No two lines that are to be resolved, or seen as being separate, shall be closer together than 1/80 the height (or hypothetical height) of the copy.*

19.7.3 Special Copy for Slides

If you make your slides, or have them made, according to the principles of Secs. 19.7.1 and 19.7.2, you will find that you cannot simply photograph the figures you have prepared for publication. You will practically always have to make up copy expressly for the slides. The curves of Fig. 8 are shown in Fig. 15 redrawn in a slide version. Notice the larger letters. (Remember that this graph has been condensed. To preserve the 1/40 relationship, the smallest letters have been made with a 120 Leroy guide. If the figure had not been condensed from 8½ x 11 inches, the smallest letters would be the 175 size.) Notice also the shortened captions, the bolder curves, the substitution of stubs for the grid. A slide made from this copy will be easily legible even in the last row of a smoke-filled auditorium.

On the other hand, if you start out by preparing copy for the slides, you can use this same material for the published version of a talk. It will look bold and black, but it will be legible and neat when it is reduced to one-column size.

19.7.4 Negative Slides

When an ordinary positive image of a drawing or graph is projected, most of the screen is very intensely illuminated. The brilliant screen in the otherwise dark room is at least unpleasant to look at; it can cause objectionable eyestrain.

The use of negative slides, which are becoming increasingly popular, avoids this undesirable effect. In the preparation of a negative slide the copy is photographed from a distance that makes the image just fit the standard slide mask opening. Then the negative is simply inserted in a slide mount.

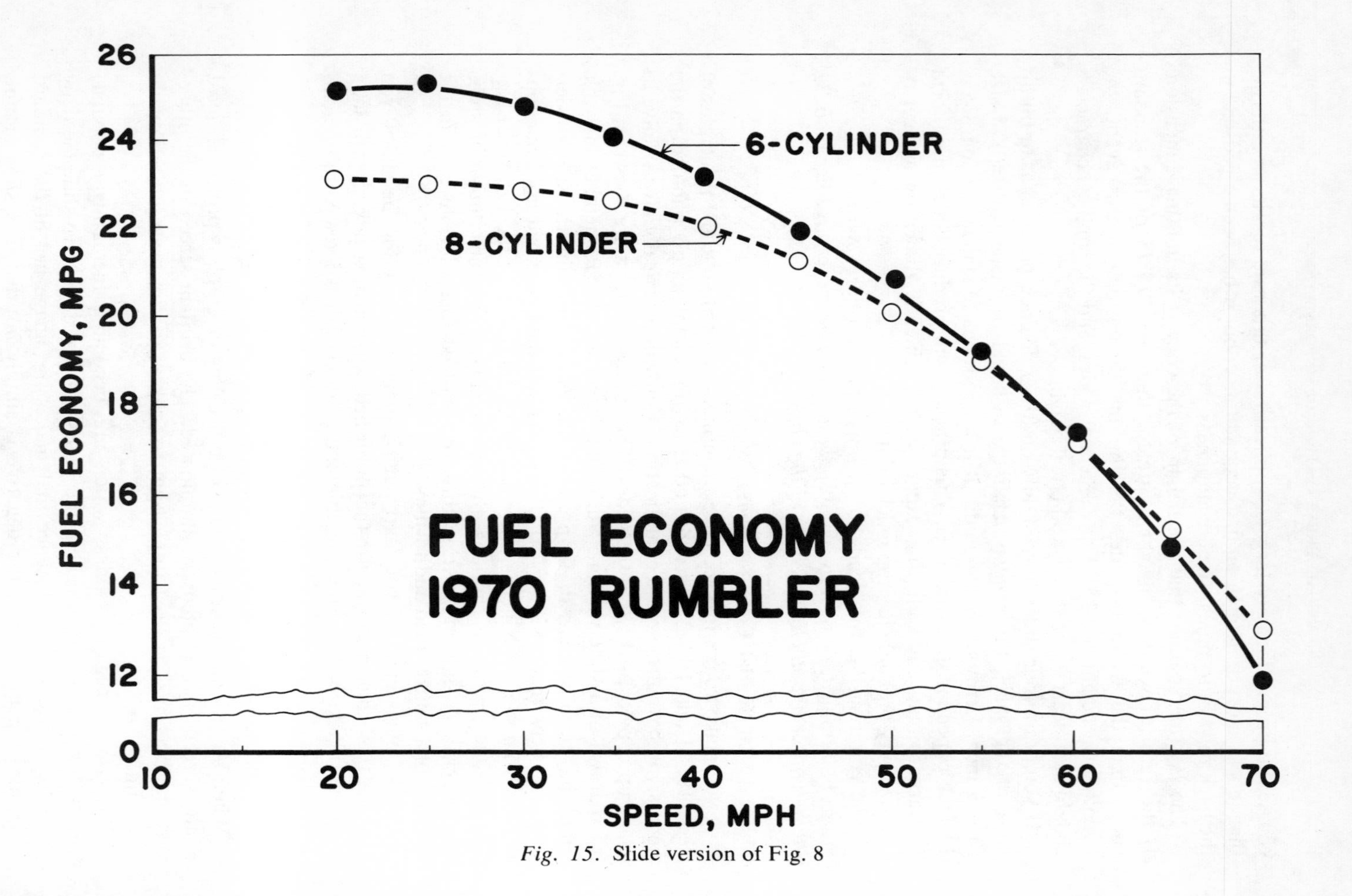

Fig. 15. Slide version of Fig. 8

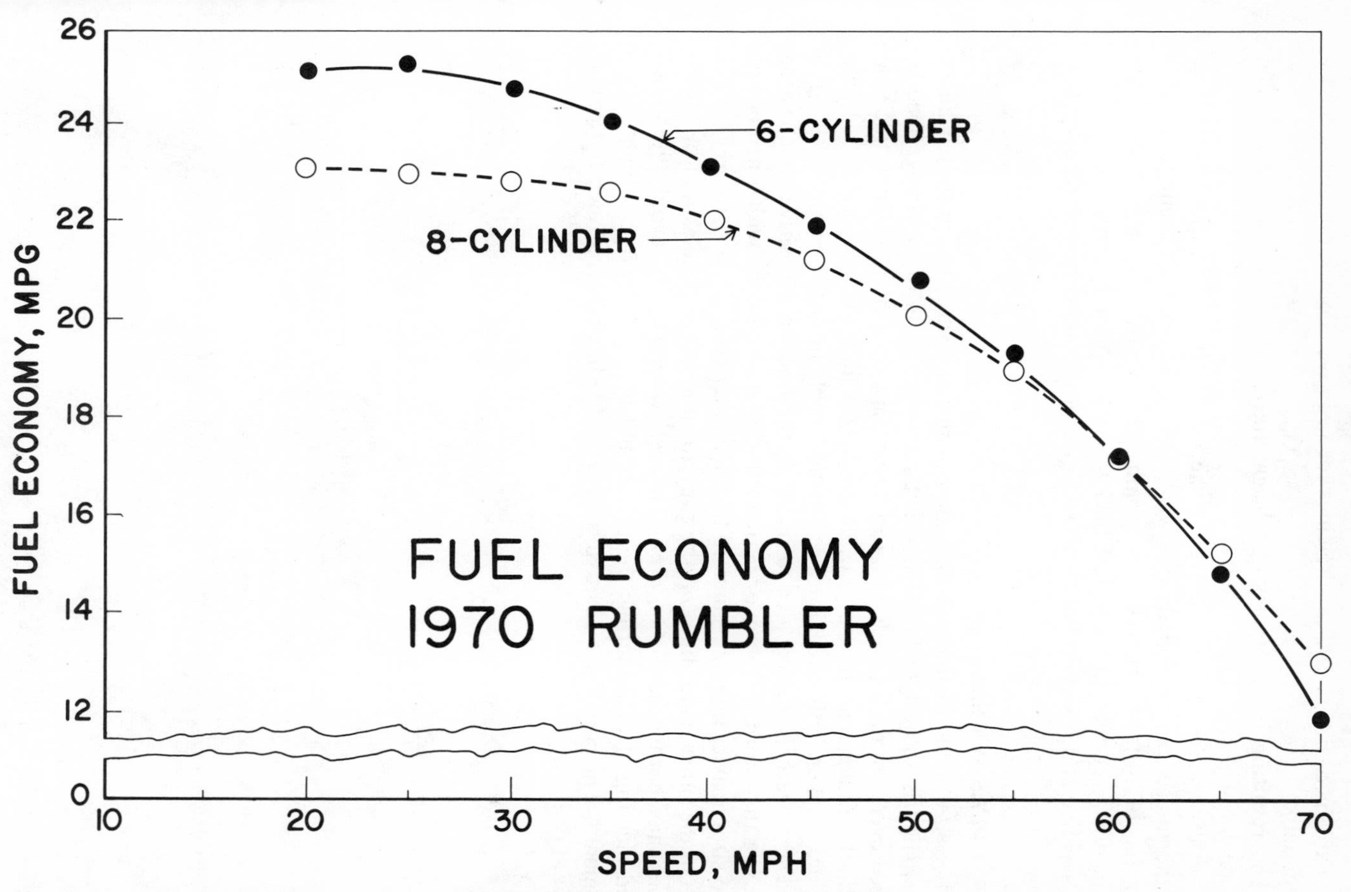

Fig. 16. Negative slide version of Fig. 15

(As explained in Sec. 19.7.5, drawings for negative slides should be specially prepared.)

The resulting image consists of bright lines on a dark background. The screen does not glare unpleasantly.

Negative slides lend themselves to the use of color. The transparent lines of the image can be tinted with water colors, so that different parts of a drawing or different curves on a graph can be easily distinguished. If the color runs over onto adjacent dark background, no harm is done.

Negative slides should not be shown in a projector without forced ventilation.

19.7.5 Pen Size

Because of the brilliant illumination of the projector, the projected image of a slide is subject to halation—that is, bright parts of the image tend to bleed over into dark parts. As a result, the dark lines on a positive slide tend to look thinner than they are, while the bright lines on a negative slide tend to look thicker. This effect is especially noticeable in the lettering.

Therefore it is a good idea to make all the lines in a positive slide a little heavier than normal. Leroy or Wrico lettering may be done with a pen one or two sizes heavier than the standard for each letter size.

On the other hand, lines on the copy for a negative slide should be a little finer than normal, *and should be kept well separated.* To avoid "filling up" and blurring of the lettering, the pen should be one or two sizes lighter than standard. The copy for a positive slide (Fig. 15) has been redrawn for a negative slide in Fig. 16.

20. In Conclusion

Whenever you have a technical reporting job to do, remember:

(1) Be as clear and as brief as you can.

(2) Design your material for your intended audience.

(3) Start with a statement of the purpose and the basic nature of the whole; then fill in the details. Orient your readers thoroughly.

(4) Don't bury important ideas under a mass of details. Eliminate unimportant details or get them out of the way in an appendix.

(5) Use headings and subheadings liberally.

(6) Be consistent.

(7) Be specific.

(8) Present opinions as opinions, not as facts. In a report, put expressions of opinion in the discussion section.

(9) Outline before you start writing.

(10) After you have completed a draft, set it aside as long as possible before you undertake the job of revision and polishing.

(11) Proofread your final copy carefully, and correct all errors, big or little.

(12) Use a logical order in describing machines, processes, and theories.

(13) Try to tailor the outline of every report to fit its specific subject and its specific purpose.

(14) Use one subject only for each memo report.

(15) When you want to give information to an individual rather than a group, try using the letter report. Use a combination of the report and the sales letter in writing a job-application letter.

(16) Unless the method you use in an investigation is more important than the results of the investigation, subordinate the description of the method to the presentation and discussion of the results. Provide only as much information about method as your readers will need.

(17) To save repetition, try to combine the presentation and discussion of results.

(18) Provide an *independent* summary—stressing results or conclusions—at the beginning of every formal report and of every informal report that will be helped by one. Whenever possible, make your summaries and abstracts informative rather than simply descriptive.

(19) Do not be afraid to draw conclusions and make recommendations if you have been authorized to and if the data warrant them.

(20) Put into the appendix any material that needs to be included in the report but that is not an integral part of the main message. Key every section of the appendix to the text by a specific reference.

(21) For any report of more than three or four pages, provide a table of contents that will serve as an outline of the report.

(22) Be sure that your thesis subject requires some real research on your part.

(23) Save the time of other people by preparing adequate interview material and questionnaires.

(24) Make instructions authoritative and easy to carry out.

(25) If possible, include explanations and precautions in instructional material.

(26) Analyze your material thoroughly before you write a proposal.

(27) Determine whether the proposal is solicited or unsolicited. The difference will be reflected in the introductory material.

(28) Near the beginning of a proposal, present your solution. Don't make the reader wade through the detailed discussion before you tell him what you intend to do.

(29) Make sure you know what the unifying idea will be before you start an article.

(30) Use article leads appropriate to the readers.

(31) Write clearly and simply, using short, plain, concrete words. Eliminate unnecessary words. Avoid involved constructions, indirect expressions. Watch out for sentences that are too long, but try to vary the length of your sentences.

(32) Write grammatically. Watch out especially for vague pronouns, dangling modifiers.

(33) Spell correctly.

(34) Punctuate properly, as a help to your readers. Test your punctua-

tion by reading aloud; a drop in your voice or a pause usually indicates the need for a comma (or a heavier mark). In particular, do not omit the comma (a) before the conjunction in a compound sentence, (b) around the nonrestrictive modifier, (c) before the conjunction in a series. Hyphenate compound adjectives that precede their nouns.

(35) Be sure your readers will know the meaning of any abbreviations you use. In scientific and technical abbreviations: (a) omit periods unless the abbreviation forms a word; (b) omit capitals unless the word abbreviated is a proper noun; (c) use the same abbreviation for singular and plural.

(36) Use Arabic numerals freely.

(37) Allow at least 1¼ inches margin on the left edge of any sheet that is going to be bound into standard report covers.

(38) Use tables, graphs, and diagrams liberally. Key every one of them to the text by an early reference. Keep them simple and uncluttered.

(39) Choose a rational scale for every graph, and mark it plainly, with units. Use the coarsest grid that will permit readings to the desired precision. Make all curves as bold as possible.

(40) When you have to deliver a talk, prepare a simple version of your material, concentrating on a full, clear presentation of only the important points. Provide frequent recapitulations. Do not memorize your speech, and do not read it unless you must. Speak slowly, clearly, and loud. Modulate your voice. Look at your audience. Don't move around unnecessarily, and be on guard against nervous mannerisms. Relax.

21. Appendix

21.1 Abbreviations for Scientific and Engineering Terms (from American National Standards Institute Publication Z10.1–1941)

These forms are recommended for readers whose familiarity with the terms used makes possible a maximum of abbreviations. For other classes of readers editors may wish to use less contracted combinations made up from this list. For example, the list gives the abbreviation of the term "feet per second" as "fps." To some readers "ft per sec" will be more easily understood.

absolute . . . abs
acre . . . spell out
acre-foot . . . acre-ft
air horsepower . . . air hp
alternating-current (as adjective) . . . a-c
ampere . . . amp
ampere-hour . . . amp-hr
amplitude, an elliptic function . . . am.
Angstrom unit . . . A
antilogarithm . . . antilog
atmosphere . . . atm
atomic weight . . . at. wt
average . . . avg
avoirdupois . . . avdp
azimuth . . . az or α

barometer . . . bar.
barrel . . . bbl
Baumé . . . Bé
boiler pressure . . . spell out
boiling point . . . bp
brake horsepower . . . bhp
brake horsepower-hour . . . bhp-hr
Brinell hardness number . . . Bhn

British thermal unit[1] . . . Btu or B
bushel . . . bu

calorie . . . cal
candle . . . c
candle-hour . . . c-hr
candlepower . . . cp
cent . . . c or ¢
center to center . . . c to c
centigram . . . cg
centiliter . . . cl
centimeter . . . cm
centimeter-gram-second (system) . . . cgs
chemical . . . chem
chemically pure . . . cp
circular . . . cir
circular mils . . . cir mils
coefficient . . . coef
cologarithm . . . colog
concentrate . . . conc
conductivity . . . cond
constant . . . const
cosecant . . . csc
cosine . . . cos
cosine of the amplitude, an elliptic function . . . cn
cotangent . . . cot
coulomb . . . spell out
counter electromotive force . . . cemf
cubic . . . cu
cubic centimeter . . . cu cm, cm^3 (liquid, meaning milliliter, ml)
cubic foot . . . cu ft
cubic feet per minute . . . cfm
cubic feet per second . . . cfs
cubic inch . . . cu in.
cubic meter . . . cu m or m^3
cubic micron . . . cu μ or cu mu or μ^3
cubic millimeter . . . cu mm or mm^3
cubic yard . . . cu yd
current density . . . spell out
cycles per second . . . spell out or c
cylinder . . . cyl

decibel . . . db
degree[2] . . . deg or °
degree centigrade . . . C
degree Fahrenheit . . . F
degree Kelvin . . . K
degree Réaumur . . . R
delta amplitude, an elliptic function . . . dn
diameter . . . diam
direct-current (as adjective) . . . d-c
dollar . . . $
dozen . . . doz
dram . . . dr

efficiency . . . eff
electric . . . elec
electromotive force . . . emf
elevation . . . el
equation . . . eq
external . . . ext

farad . . . spell out or f
feet per minute . . . fpm
feet per second . . . fps
fluid . . . fl
foot . . . ft
foot-candle . . . ft-c
foot-Lambert . . . ft-L
foot-pound . . . ft-lb
foot-pound-second (system) . . . fps
foot-second (see cubic feet per second)
free on board . . . fob
freezing point . . . fp

[1] Abbreviation recommended by the A.S.M.E. Power Test Codes Committee. B = 1 Btu, kB = 1000 Btu, mB = 1,000,000 Btu. The A.S.H.&V.E. recommends the use of Mb = 1000 Btu and Mbh = 1000 Btu per hr.

[2] There are circumstances under which one or the other of these forms is preferred. In general the sign ° is used where space conditions make it necessary, as in tabular matter, and when abbreviations are cumbersome, as in some angular measurements, i.e., 59° 23′ 42″. In the interest of simplicity and clarity the Committee has recommended that the abbreviation for the temperature scale, F, C, K, etc., always be included in expressions for numerical temperatures, but, wherever feasible, the abbreviation for "degree" be omitted; as 69 F.

frequency . . . spell out
fusion point . . . fnp

gallon . . . gal
gallons per minute . . . gpm
gallons per second . . . gps
grain . . . spell out
gram . . . g
gram-calorie . . . g-cal
greatest common divisor . . . gcd

haversine . . . hav
henry . . . h
high-pressure (adjective) . . . h-p
horsepower . . . hp
horsepower-hour . . . hp-hr
hour . . . hr
hour (in astronomical tables) . . . h
hundred . . . C
hundredweight (112 lb) . . . cwt
hyperbolic cosine . . . cosh
hyperbolic sine . . . sinh
hyperbolic tangent . . . tanh

inch . . . in.
inch-pound . . . in-lb
inches per second . . . ips
indicated horsepower . . . ihp
indicated horsepower-hour . . . ihp-hr
inside diameter . . . ID
intermediate-pressure (adjective) . . . i-p
internal . . . int

joule . . . j

kilocalorie . . . kcal
kilocycles per second . . . kc
kilogram . . . kg
kilogram-calorie . . . kg-cal
kilogram-meter . . . kg-m
kilograms per cubic meter . . . kg per cu m or kg/m^3
kilograms per second . . . kgps
kiloliter . . . kl
kilometer . . . km
kilometers per second . . . kmps
kilovolt . . . kv
kilovolt-ampere . . . kva
kilowatt . . . kw
kilowatthour . . . kwhr

lambert . . . L
latitude . . . lat or ϕ
least common multiple . . . lcm
linear foot . . . lin ft
liquid . . . liq
liter . . . l
logarithm (common) . . . log
logarithm (natural) . . . $\log_e$ or ln
longitude . . . long. or λ
low-pressure (as adjective) . . . l-p
lumen . . . l
lumen-hour . . . l-hr
lumens per watt . . . lpw

mass . . . spell out
mathematics (ical) . . . math
maximum . . . max
mean effective pressure . . . mep
mean horizontal candlepower . . . mhcp
megacycle . . . spell out
megohm . . . spell out
melting point . . . mp
meter . . . m
meter-kilogram . . . m-kg
mho . . . spell out
microampere . . . μa or mu a
microfarad . . . μf
microinch . . . μin.
micromicrofarad . . . $\mu\mu$f
micromicron . . . $\mu\mu$ or mu mu
micron . . . μ or mu
microvolt . . . μv
microwatt . . . μw or mu w
mile . . . spell out
miles per hour . . . mph
miles per hour per second . . . mphps
milliampere . . . ma
milligram . . . mg
millihenry . . . mh
millilambert . . . mL
milliliter . . . ml
millimeter . . . mm
millimicron . . . mμ or m mu

million .... spell out
million gallons per day .... mgd
millivolt .... mv
minimum .... min
minute .... min
minute (angular measure) .... ′
minute (time) (in astronomical tables) .... m
mole .... spell out
molecular weight .... mol. wt
month .... spell out

National Electrical Code .... NEC

ohm .... spell out or Ω
ohm-centimeter .... ohm-cm
ounce .... oz
ounce-foot .... oz-ft
ounce-inch .... oz-in.
outside diameter .... OD

parts per million .... ppm
peck .... pk
penny (pence) .... d
pennyweight .... dwt
per .... [See Sec. 17.1.2]
pint .... pt
potential .... spell out
potential difference .... spell out
pound .... lb
pound-foot .... lb-ft
pound-inch .... lb-in.
pound sterling .... £
pounds per brake horse-power-hour .... lb per bhp-hr
pounds per cubic foot .... lb per cu ft
pounds per square foot .... psf
pounds per square inch .... psi
pounds per square inch absolute .... psia
power factor .... spell out or pf

quart .... qt

radian .... spell out
reactive kilovolt-ampere .... kvar
reactive volt-ampere .... var
revolutions per minute .... rpm
revolutions per second .... rps
rod .... spell out
root mean square .... rms

secant .... sec
second .... sec
second (angular measure) .... ″
second-foot (see cubic feet per second)
second (time) (in astronomical tables) .... s
shaft horsepower .... shp
sine .... sin
sine of the amplitude, an elliptic function .... sn
specific gravity .... sp gr
specific heat .... sp ht
spherical candle power .... scp
square .... sq
square centimeter .... sq cm or cm^2
square foot .... sq ft
square inch .... sq in.
square kilometer .... sq km or km^2
square meter .... sq m or m^2
square micron .... sq μ or sq mu or μ^2
square millimeter .... sq mm or mm^2
square root of mean square .... rms
standard .... std

tangent .... tan
temperature .... temp
tensile strength .... ts
thousand .... M
thousand foot-pounds .... kip-ft
thousand pound .... kip
ton .... spell out
ton-mile .... spell out

versed sine .... vers
volt v
volt-ampere .... va
volt-coulomb .... spell out

watt .... w
watthour .... whr
watts per candle .... wpc
week .... spell out
weight .... wt

yard .... yd
year .... yr

21.2 Specimens

Section 21.2 contains specimens that illustrate the ideas discussed in *Technical Reporting*. Remember that these are simply examples from a wide choice of acceptable forms.

The originals of most of these specimens were typed on 8½ x 11-inch paper. The condensation required to make them fit the pages of this book has resulted in some crowding, but the specimens will serve their illustrative purpose.

Reports and letters of this kind are usually typed on only one side of the paper. This practice has been largely followed here: most of the specimen material appears on the right-hand pages only. On the facing left-hand pages are pertinent remarks and comments.

21.2.1 Specimen Memo Report

This piece of writing is a good illustration of the usual brevity of the memo report.

It wastes no time in dealing with those items that can be standardized, such as the report subject, from whom it came, and to whom it is directed. It uses headings and listings and attempts to put information within each section in parallel construction.

In content, this memo report is divided into three parts: the assignment from the director of the laboratory, the requirements, and the conclusions and recommendations. If it were necessary, other sections might be added. But in any event, the section on conclusions and recommendations would probably be found early in the report.

21.2.1 SPECIMEN MEMO REPORT

Inter-Office Correspondence

LOGAN HYDRAULIC CORPORATION
RIPLEY, NEW MEXICO

To: Mr. Lewis Forman
Chief Engineer

cc: Mr. W.M. O'Hara
Director, R&D

From: W.A. Porter; H.T. Gray; S. Fenil

Date: March 25, 1970

Subject: CRYOGENIC LABORATORY FOR R&D

At the request of Mr. W.M. O'Hara, the possibility of establishing a cryogenic laboratory has been investigated. We were directed to make a survey of available facilities before determining laboratory costs and to make recommendations.

Requirements

1. The current estimate of necessary laboratory space is approximately 500 square feet. This space should be provided with sprinklers, heating (preferably hot water or steam), lighting, power, and ventilation.

2. Work benches should be metal.

3. Materials to be used should be noncombustible.

4. The ventilation system should provide for one complete change of air every 2 or 3 minutes in a volume of about 5000 cubic feet.

Conclusions and Recommendations

1. The first space surveyed is now occupied by the Hot Hydraulic Fluids Laboratory. This space will be available when the laboratory moves into the new R&D building.

Too great an expense would be involved in modifying this facility for the Cryogenics Laboratory. This space is considerably larger than necessary and is equipped with wiring that would have to be made explosion-proof.

2. The second area surveyed is the test cells in Building 55. This space is now used for storage by the Facilities Department.

We recommend that one of these cells be used for the Cryogenics Laboratory. It is more nearly the correct size, it is already equipped with vent ducts, and most of the lighting fixtures are explosion-proof. Particular cells run along the east wall and the northeast corner of the building and are readily accessible to the interior loading dock used to receive shipments of cryogenic liquid.

21.2.2 Specimen Memorandum

This memorandum, like the preceding report, uses one of the acceptable heading forms. This heading enables you to identify the report and its subject quickly and easily.

Although it is less than a page long, this memorandum is clarified by its two subheads and its two tabulated items.

21.2.2 SPECIMEN MEMORANDUM

2345
Memorandum A-2000

Page 1 of 1

Computer Applications Project
Massachusetts Institute of Technology
Cambridge, Massachusetts

SUBJECT: HEADINGS AND NUMBERS FOR 2345 REPORTS

To: All Secretaries; Applications Group
From: J. S. Underwood
Date: May 4, 1970

Headings

As shown above, headings for informal reports on the Computer Applications Project, CID Project 2345, will differ in two respects from the regular XYZ Computer Project headings:

1. The number 2345 will replace the number 1234 in the upper left corner.
2. The first line of the title will read Computer Applications Project instead of XYZ Computer Project.

Numbers

Reports on the Computer Applications Project will be issued in the appropriate R-, E-, and A-series, but their numbers will be in a block separate from those of the XYZ Computer Project, starting with No. 2000. Numbers will be assigned by the Librarian according to the usual procedure.

Signed *J. S. Underwood*
J. S. Underwood

Approved *L. S. Thurston*
L. S. Thurston

JSU:jz

21.2.3 Specimen Survey Report

This memo is an example of a survey report. It was prepared by a student to precede the writing of an article.

You will notice that the author gives as early as possible the broad, general conclusion that the magazine might be interested in his article topic. After the general conclusion has been established, details of the survey are given; they will be useful for both the student and the teacher who will be reading the article before it is sent to the magazine.

21.2.3 SPECIMEN SURVEY REPORT

SURVEY REPORT OF

The General Engineering Review

TO: Professor L. M. Herron

FROM: Leighton D. Myrick

DATE: May 10, 1970

SUBJECT: Survey of the General Engineering Review

Purpose

I propose to write an article on the rehabilitation of old buildings, particularly in reference to urban planning and renewal. Therefore, I conducted a survey of several publications, among them the General Engineering Review, a weekly magazine published by the General Engineering Company, to determine the suitability of the material.

Procedure

I followed this plan:

1) I surveyed issues for three months.

2) I investigated each issue for editorial policy, publication and circulation, article content, article characteristics, staff makeup, and general aims and purposes.

3) I obtained style sheets and general specifications from the magazine and surveyed them.

Conclusions

I decided that an article on the rehabilitation of old buildings would be within the specifications of the General Engineering Review. I therefore plan to go ahead with the project, to be completed within six weeks.

Results of the Survey

A detailed analysis of the issues surveyed brought out the following information.

Type

Of the three general forms of technical magazines - professional, commercial, and company - the General Engineering Review is a company magazine with a style and format similar to commercial magazines. Like most commercial magazines, it gives an honorarium of $40-$60 a printed page.

You will benefit by looking over this survey report carefully to see how the headings are handled, how listings are used, and how various items have been grouped as logically as possible.

Aims

The Review in general serves two purposes: (1) to describe general accomplishments in science and engineering and (2) to describe specific accomplishments within the company.

Publication and Circulation

The editorial staff is composed of the managing editor, two research and engineering editors, one copy editor, and an editorial advisory council.

The council is made up of three members, each an authority in his chosen field of science or engineering.

Editorial Policy

The policy follows the general aims already described. Each issue contains the following:

(1) A theme for the issue. This may be some particular branch of engineering or a specific theme occasioned by a national conference or similar event.

(2) An editorial on the week's issue. Occasionally the editorial is written by a guest author who may or may not be employed by the company.

Other Features

(1) Articles are about evenly divided between those that are staff prepared and those from outside sources.

(2) Once a month the Review contains an article on a continuing series, "Engineering Societies of America." This is usually written by an official of the society.

Article Content

The articles deal with topics of current interest. Occasionally a historical subject is included, but only if it has some timely interest.

The Review seems to prefer articles with a strong applied emphasis. Rarely does it accept a largely theoretical subject. In other words, each article has a practical application, particularly for readers with a strong engineering background.

Specifications

An author's guide supplied by the Review lists these specifications:

(1) The style must be semitechnical, capable of being understood by a wide range of engineering and science readers.

(2) Each article must be accompanied by several illustrations or other graphic aids.

(3) The length should be between 2,000 and 3,000 words.

(4) Queries should be made in advance.

(5) Articles of an idustrial nature should be accompanied by a release or an endorsement by the company employing the author.

21.2.4 Specimen Survey Report

Here is a survey report as it was prepared in a division of a large engineering company. The names of people and places have been changed, but the report itself is the same as originally written.

We are not going to analyze this report for you. See how well written you think it is and how well it follows the principles for this kind of report.

21.2.4 SPECIMEN SURVEY REPORT

JONES MANUFACTURING COMPANY

Philadelphia, Pennsylvania

May 29, 1970

To: Mr. R. M. Barth, General Manager

From: Mr. J. P. Howds
Production Department
General Offices

Subject: PRODUCTION AT PLANT NO. 6, SCRANTON, PA.

As requested in your letter of May 15, an investigation has been made of the production work at our Scranton plant. A report on it is now submitted.

Procedure

The survey was divided into three parts: plant, personnel, and methods. An analysis of each of these is included, together with specific recommenda--tions.

PLANT

The Scranton plant, No. 6, manufacturing washing machines, consists of a factory building, a boiler house, and a garage. The factory building and boiler house are 15 years old; the garage 5. The factory building, constructed of re-inforced concrete, has two stories and about 10,000 sq ft of floor space. The boiler house is a small, one-story brick building containing a boiler used for heating the factory and an air compressor supplying compressed air for various processes. The garage is of steel-frame construction with stalls for 8 cars or trucks; a small room at one end contains oil, grease, and a small assortment of tools.

Factory Building

Heating. Though there was no heat in the building at the time of the

2

the investigation, Superintendent Sullivan states that the system works efficiently.

Lighting. Except for requirements of the milling machines, the lighting is generally satisfactory. Electric lamps hanging from the ceiling are enclosed in translucent globes. Walls and ceilings are painted white.

It is recommended that each milling machine should be individually lighted because of the highly accurate work required.

Ventilation. Excessive heat at times results in complaints and a perceptible slackening in production. Ventilation is provided by windows and by ventilators in the roof, but at the time of the inspection the temperature was 80° -- too high for efficient work.

It is recommended that this situation be relieved by the installation of oscillating electric fans on the side walls at intervals of 50 ft, except in the storeroom and shipping room, where the temperature should be normal.

Power. Power is supplied by the local electric company. No changes are recommended.

Machinery. Though most machinery is found to be in serviceable condition, some replacements are needed. Table 1 indicates machines that should be junked and the number needed to replace them.

Table 1

Machinery Recommended to be Junked

Machines	No. to Be Junked	No. to Be Replaced
Brown & Sharpe Grinder	2	1
Cleveland Drill Press	1	1
South Bend Lathe	2	1

Where the table shows fewer replacements than machines to be junked, a time study has indicated that production would not be decreased as a result of partial replacements.

Maintenance. No regular systems of inspection and lubrication are in force. The maintenance force gives attention only to actual breakages. Lubrication seems to be done without regard to the special requirements of different machines.

It is recommended that a system of weekly inspections of each piece of equipment be established.

It is also recommended that a definite lubrication schedule be established with attention given to the specific requirements of each machine.

These precautionary measures, which should decrease breakages and replacement costs, are possible without adding to the present maintenance force.

Boiler House

No changes are recommended.

Garage

No changes are recommended.

PERSONNEL

Present Working Force

The working force at the Scranton plant totals 520, of whom 50 are women. About half may be classified as skilled. In general, labor relations have been satisfactory, and turnover has been somewhat lower than in other plants. Table 2 shows distribution by departments.

Possible Reductions

By applying more efficient methods, as discussed in the section on "Production Standards," it should be possible to reduce the working force as shown in Table 2.

Table 2

Possible Reductions in Working Force

Department	Present Force	Reduced Force
Stores	8	7
Shipping	10	9
Machine	100	95
Stamping	25	23
Pressing	50	48
Painting	10	10
Inspection	75	70
Assembly	150	146
Testing	40	40
Maintenance	5	5
Tool	15	15
Engineering	2	2
Office	15	15
Transportation	15	15
Total	520	500
Reduction		20

It is also recommended that the management undertake to weed out gradually all workers whose production records for a period of six months do not reach 60 percent of the standard set for the particular jobs they are doing.

METHODS

Production Standards

Time studies indicate that a substantial increase in production is possible if fair standards are set. These time studies were made of workers selected by their foremen as about the average. Table 3 shows the differences between the standard times set and the actual working times during the study:

4

Table 3

Standard Times
(Minutes)

Job	Actual Time	Standard Time
Cutting ring gear	7	5
Turning drive shaft	12	10
Painting washer frames	3	2 ½
Testing wringer drive	8	6
Assembling drive gears	10	7

Table 3 seems to indicate that there is sufficient room for improvement to justify a more extensive program of standard-setting. An increase of 5 percent in production should be a fair estimate of the results.

It was on the basis of the results shown in Table 3 that the schedule of possible reductions in working force shown in Table 2 was devised.

Wage Payments

All employees, except office and storeroom workers and executives above the grade of foreman, were found to be paid on a piece-rate basis.

With the establishment of standard times for most jobs, it is recommended that employees now on piece rate be placed on a bonus system, since such a change ought to stimulate production.

Production Control

The lack of production control leaves the management out of touch with much that goes on in the factory. It is not possible to know at all times how work is progressing or why delays occur.

It is recommended that a system similar to the Gantt chart be instituted.

SUMMARY OF RECOMMENDATIONS

Factory Building

1. Each milling machine should be individually lighted.

2. Electric fans should be installed except in storeroom and shipping room.

3. Individual machines (see Table 1) should be replaced.

4. Each machine should be inspected weekly.

5. A lubrication schedule should be established.

Personnel

1. The working force (see Table 2) should be reduced.

2. Unproductive workers should be weeded out.

Methods

1. An attempt should be made to increase production by 5 percent.
2. Piece-rate system should be replaced by a bonus system.
3. A production-control system should be instituted.
4. Stored material should be arranged.
5. A perpetual inventory system of stored material should be established.
6. Methods of shipping room should be changed similar to those in storeroom.

21.2.5 Specimen Progress Report

This report, obtained from a well-known engineering company, illustrates a short progress report for internal reading. It follows the form for most memo reports. The project is identified, the period covered is stated, completed work is recorded, and work in progress is discussed.

21.2.5 SPECIMEN PROGRESS REPORT

LOGAN HYDRAULIC CORPORATION
Ripley, New Mexico

January 14, 1970

TO: Mr. Lewis Forman
Chief Engineer

FROM: B. L. McGovern

PERIOD: June 15 - December 15, 1969

SUBJECT: Integrated Antenna - Radome Facility Progress Report

A. FACILITY STATUS

1. Construction of the facility has been completed except for refinements now in progress. These refinements do not interfere with the operation of the facility.

2. Storage area for electronic test equipment and parts was completed during the past month.

3. The installation and check-out and the calibration of test equipment for both the radome and antenna test ranges were completed on December 8.

B. WORK IN PROGRESS

Tests involving the F-105D radome have been conducted during the past two months. Summary of tests to date is as follows:

1. An investigation study of the false target return of the NASARR Radar was carried out. This study involved the radar antenna measurement with and without the radome. The study was carried out as a joint effort of both RAC and the radome vendor.

2. At present the radome facility at GHC is being used to back up the vendor by performing the electrical tests on production radomes. These tests involve pattern boresight and transmission tests. At the same time final boresight correction is incorporated in each radome tested.

21.2.6 Specimen Service Report in Memo Form

This memo is an example of a trip report. It originated in a government office and was used only for internal circulation. It is a good example of how a specific agency has adapted certain principles of reporting to a basic situation occurring over and over again.

The first page only has been reproduced here, as the minutes referred to in 4 would take up too much space. However, by looking over the section on results, you can tell what business was taken up, just as the original reader was able to.

21.2.6 SPECIMEN SERVICE REPORT IN MEMO FORM

OFFICE MEMORANDUM

LUMMUS ARSENAL
10 August 1970

TO: Chief, National Commercial Division
THRU: Mr. A. M. Farrar

FROM: Henry S. Bowers, Publications Section
Industrial Engineering Branch

SUBJECT: Report of Travel

PLACE VISITED AND DATE:
Ordnance Weapons Command, Red Island Arsenal,
Red Island, Ohio, 7 August 1970

PURPOSE OF VISIT:
To attend Second Session of Task Group,
Inspection Engineering Handbook Conference.

PERSONNEL PRESENT

L. M. Brown	Ord. Ammunition
H. D. Scanlon	Ord. Tank Auto
R. Vincent	Waterville Arsenal
E. T. Vicuma	Aberdeen Arsenal
T. Andrews	Red Island Arsenal
L. P. Potter	Ord. Weapons Command
H. S. Bowers	Lummus Arsenal

RESULTS ACCOMPLISHED AND REMARKS:

1. Meeting opened by L. M. Brown, Chairman, at 0900 hours.
2. Next meeting of the group set for 19 September 1970 at Aberdeen Arsenal.
3. Agenda of the meeting was as follows:
   a. Introduction of Project Engineer for Handbook Contract.
   b. Completion of work on procedural outlines for various procurement patterns.
   c. Current handling and standard phraseology concerning how the Agent of Inspection shall be designated in specifications.
4. Minutes of the meeting are attached.
5. A course on the preparation of Military Specifications will be given at Aberdeen Arsenal. The first class is scheduled for 22 September 1970. Attendance by Lummus Arsenal representatives is recommended.

21.2.7 Specimen Trip Report

Here is another trip report, this time from a state health department. Reports of this kind are used by the agency for its annual report and therefore are important as records of the preceding year.

The content of this report is representative of the many conferences and field trips made by members of this government agency. Analyze this report to see how an exploratory meeting with few tangible results has been organized into a coherent piece of information.

21.2.7 SPECIMEN TRIP REPORT

STATE DEPARTMENT OF SANITATION

Field Trip Report

December 17, 1969

TO: Mrs. Charlotte P. Ogilvie
Supervisor

FROM: Lewis H. Sardon
Health Consultant

1. Date: December 12, 1969

2. Location and Agency: City of Talmot, Common Council Meeting

3. Purpose of Trip: To record questions of the Common Council concerning the proposed sewage construction project and to show the film The Time Is Now.

4. Persons Seen: Common Council Members, Mayor and President of Common Council, Merrick County Health Commissioner and Engineer, and City of Talmot Consulting Engineer.

Summary:

1. The meeting was conducted in the form of a symposium. Those speaking were: Dr. R. M. Kandell; Richard Harvey, Merrick County Health Department; Henry Seymour, sanitary engineer, N.Y.S.H.D.; and Robert Farmer, consulting engineer. They spoke to the City Council and explained the proposed sewage construction project.

2. Approximately 40 people attended to hear the background and history of sewage pollution and the efforts to solve the problem. They were given an informational brochure on the subject.

3. The meeting was considered a start in the right direction, but the method of fixing sewer charges was confusing to many. An attempt at clarification will be made in the local paper.

Follow-up:

Answers to the questions by the Common Council will be prepared and submitted for distribution to the members of the council.

21.2.8 Specimen Letter Report

This specimen of a letter report combines the features of the typical business letter and the informal report.

You will notice that it has been written by one company to another. The mechanics of the business letter have been observed: letterhead, date, inside address, salutation, and complimentary close. These items have been put in their conventional places.

But this specimen is primarily a report. A survey has been made, certain results have been achieved, and recommendations are suggested. The information could have been put in straight report form with title page, summary, and conclusions. The writer, however, showed good judgment in the form he chose. He undoubtedly felt that the letter gave a more personal tone to his report.

21.2.8 SPECIMEN LETTER REPORT

MOHAWK AUTOMATIC SIGNALS, INC.
Rensselaer, New York

January 6, 1970

Mr. P. O. Laughlin
Superintendent of Telegraph
New York Union Railways, Inc.
Albany, New York

Subject: Crossing Protection at Lawrence Avenue, Albany, New York

Dear Mr. Laughlin:

In your letter of December 13, 1969, you stated that poor crossing conditions existed at Lawrence Avenue, Albany.

You asked this company to make a survey of the crossing conditions, with the possibility of the company providing visible and audible warning for both northward and southward train movement.

Results and Recommendation

We have completed the survey, for which specific data are appended. The results of the survey strongly indicate that serious traffic congestion exists and that a warning system should be provided.

We recommend that an annunciator system be installed at the Lawrence Avenue crossing. Such an installation would provide warning in these directions:

1. Southward on either of your main tracks for approximately 1600 feet
2. Northward on either of your main tracks for approximately 1350 feet
3. Northward on the Northern Belt Line track for approximately 1000 feet

Estimate of Costs

Our estimate to install annunciators is as follows:

Material..................................	$1545
Freight and stores........................	210
Labor.....................................	735
Engineering and contingencies.............	175
Total	$2665

Installation Requirements

Your speed limit in this territory is now 20 mph for main track movements and 15 mph for movements to and from the Northern Belt Line.

If we contract to install the annunciator system, we will request that train movements over the crossing only be further reduced while installation is being made.

We will further request that during the installation the flagman be provided with a large Stop banner surfaced with reflex-reflecting sheet material.

Sincerely yours,

Murray O. Harris

Murray O. Harris
Superintendent, Signal Division

21.2.9 Specimen Letter Requesting Information

This letter is set up in a form that is widely used for business correspondence. (Many organizations do not indent for paragraphs, but simply double-space between paragraphs.)

Notice that the tone of this letter is personal and cordial. The writer has realized that he is asking a favor; therefore he has used two basic principles for a letter of this type—he has explained why he needs the information, and he has definitely made the letter easy to answer by asking definite questions.

The language is the same as the language you would use in a report—simple, straightforward, and matter-of-fact, with no extra embellishment.

21.2.9 SPECIMEN LETTER REQUESTING INFORMATION

December 20, 1969

Elbert Photographic Supply Company
29 South Oak Street
Boston, Massachusetts 02109

Gentlemen:

In connection with a paper I am preparing on the production of photographic slides, I should like to ask your assistance. I am a senior in chemical engineering at MIT.

So far I have not been able to get from any other source pertinent information on the coloring of negative slides. Specifically, I would like to know the steps in the process, the solutions used, and any particular techniques which have proved valuable.

I hope that I am not taking up too much of your time with my questions. I will greatly appreciate your help, and I will be very glad to send you a copy of my completed paper if you would care to look it over.

Sincerely,

Fred L. Johanssen

Fred L. Johanssen

21.2.10 Specimen Letter Giving Information

This letter is a reply to the letter of Sec. 21.2.9. Because the student made his request in readable form, the firm has been able to supply the information quickly and easily.

Note that the writer refers to the previous correspondence and that he has reacted happily to the offer made by the student. That offer helped to put the two of them on common ground. The information given is listed and numbered. The letter ends as cordially as it began.

21.2.10 SPECIMEN LETTER GIVING INFORMATION

ELBERT PHOTOGRAPHIC SUPPLY CO.

29 South Oak Street
Boston 20, Massachusetts

December 29, 1969

Mr. Fred L. Johanssen
Massachusetts Institute of Technology
Cambridge, Massachusetts 02139

Dear Mr. Johanssen:

We are glad to provide you with information about the coloring of negative slides. Thank you for your offer; we should like to see a copy of your completed paper.

We find that the coloring process is done best in these three steps:

1. Apply a water-insoluble lacquer immediately around the lines or letters to be colored, to act as a dam.

2. After the lacquer has dried, paint the enclosed area with a water solution of an aniline dye.

3. After the dye has dried thoroughly, remove the lacquer with a suitable solvent.

If other colors are to be applied to the same slide, the process is repeated, a new lacquer dam being provided for each color. If the color is too dense, any excess can be removed with a damp cloth.

If you need any further information we shall be glad to supply it, or we could give you a demonstration of the process the next time we have an order for colored negative slides.

Very truly yours,
ELBERT PHOTOGRAPHIC SUPPLY CO.

Victor O. Federer

Victor O. Federer, Vice President

VOF:ba

21.2.11 Specimen Letter Requesting Information

A second letter requesting information is also written in a personal tone. However, this correspondence—this letter and the reply following it—is between two professional men about a professional subject.

Consequently, not so much time is spent on the reason for making the request, and the letter in general is short and to the point.

21.2.11 SPECIMEN LETTER REQUESTING INFORMATION

UNITED STATES DIVISION OF ECONOMIC FORECASTING

Washington, D.C.

March 15, 1970

Mr. Angus Macmillan, Director
Research and Development
Employment Security Division
Labor Department
Washington, D.C. 20575

Dear Mr. Macmillan:

I have been able to obtain a copy of your publication Economic Indicators and have found a number of items in it that could be applied to the work of this division. Would it be possible to be placed on your mailing list on a regular basis?

I'm sure that you will agree that there should be more exchange of ideas on matters pertaining to our respective departments. For example, I am interested in any economic time series available and in any other similar studies that would be useful to a new program we have recently started.

If at any time I can repay this favor, I would be glad to do so.

Sincerely,

Ralph R. Peters

Ralph R. Peters
Principal Economist

21.2.12 Specimen Letter Giving Information

This second letter giving information has fewer report characteristics than the preceding one, yet it accomplishes its purpose quickly and clearly.

A subject line identifies the original request. This is followed by a statement of the action taken; thus, the reader knows immediately what to expect from his request. Each paragraph deals with a section of the information requested.

The tone of the letter is good, helped particularly by the courteous ending.

21.2.12 SPECIMEN LETTER GIVING INFORMATION

Employment Security Division

LABOR DEPARTMENT

March 28, 1970

Mr. Ralph R. Peters
Principal Economist
United States Division of
Economic Forecasting
Washington, D.C. 20575

Dear Mr. Peters:

Subject: Economic Indicator Program

We are enclosing several copies of our publication Economic Indicators. This is in answer to your request of March 15. We have placed your unit on our mailing list and you will receive our publication monthly.

At present this department publishes fourteen economic time series. The choice for publishing these was based on availability, significance, source reliability, and behavior of data. All data are seasonally adjusted in accordance with the procedure set up in the BLS Seasonal Factor Method (1964).

Presentation of the material in Economic Indicators follows closely the format used by the United States Department of Commerce, Bureau of the Census. We have enclosed a set of Explanatory Notes for Economic Indicators and a suggested method of charting deseasonalized data.

Please feel free to write us for additional information as your program develops.

Sincerely yours,

Angus Macmillan

Angus Macmillan, Director
Research and Development

21.2.13 Specimen Job-application Letter (Straight Letter)

It is customary to put the return address to the right of center on paper that does not have a printed letterhead.

Notice that the letter begins by orienting the reader—telling him what it is about. Identification is made easy by reference to the name of a person known to both writer and reader. The writer then tells why he particularly wants to work for this company.

These preliminaries taken care of, the letter describes the writer, beginning with his educational background and practical experience.

Following these statements are some of the vital statistics so often put on a separate data sheet (see the specimen in Sec. 21.2.14). Particularly important are the honors the writer has received during his years at college. The letter closes with a request for an interview, which is usually the first step in obtaining a job.

The writer of this letter has used the first person throughout. This approach is necessary because, after all, the product he is attempting to sell is himself and his education. The tone of this letter is good, striking a nice balance between overconfidence and undue humility.

21.2.13 SPECIMEN JOB-APPLICATION LETTER (STRAIGHT LETTER)

3505 Manfred Drive
College Park, Pennsylvania 16802
February 10, 1970

Mr. Arnold S. Sutton
Division Engineer
Geophysical Services, Inc.
Littleton, Texas 79339

Dear Mr. Sutton:

At the suggestion of Dr. G. R. Fried of the Pennsylvania Institute of Technology, I am writing to apply for a position as a geophysical engineer in your company. Dr. Fried has told me that he was on the program with you at the recent conference of the American Petrological Society.

My interest in your company was stimulated by an article by one of your engineers in the October issue of the Gas and Oil Journal. In that article, Mr. Ramroth described the opportunities for young engineers both in this country and in the Middle East.

I am a senior in geophysics at the Pennsylvania Technical Institute, where I shall receive a B.S. degree early in June. I would be able to start work after June 15. I have been in the upper quarter of my class during my four years in college. During my junior and senior years I have been on a tuition scholarship. In the summer of 1968, I was one of 25 students in the country selected to work at the Alert Weather Station in Greenland. This experience was valuable for my later studies.

Besides my technical training, I have had several summers of practical experience. I worked for two summers as a greenhouse grower and one summer as assistant to the editor of the Journal of Petrology, published by P.I.T. About 70 percent of my college expenses came from part-time and summer jobs.

From the fall of 1964 to the fall of 1966, I was in the United States Army. During this time I was assigned to personnel work. I was glad to return to college to complete my education. During my senior year I have been assisting the laboratory technicians in setting up laboratory experiments and I have supervised sophomore students on field trips.

I have been the secretary of the student chapter of the National Geophysical Society; features editor of the P.I.T. Engineer; and junior class representative of the Student Council.

I am 26 years old, married, and the father of a 6-month-old daughter.

I can provide references at your request, among them Dr. Fried and Dr. L. D. Sawyer, Chairman of the Department of Geology.

I shall be glad to come to Littleton for an interview. The P.I.T. Placement Office has informed me that representatives of your company will be on campus on March 2. I have arranged to have an interview with them on that date.

Sincerely,

Ralph J. Carrier

Ralph J. Carrier

21.2.14 Specimen Job-application Letter (Data Sheet and Covering Letter)

This letter and data sheet deals with the same material as the preceding letter. But the letter is considerably shorter because much of the information is tabulated in the accompanying data sheet.

Notice, however, that the writer has not depended entirely on the data sheet. He has used the letter to point out some features in the data sheet that make him different from every other applicant. In other words, the product —the writer's services—is described in the data sheet; the letter is very definitely a sales letter drawing attention to those services.

21.2.14 SPECIMEN JOB-APPLICATION LETTER (DATA SHEET, LETTER)

3505 Manfred Drive
College Park, Pennsylvania 16802
February 10, 1970

Mr. Arnold S. Sutton
Division Engineer
Geophysical Services, Inc.
Littleton, Texas 79339

Dear Mr. Sutton:

At the suggestion of Dr. G. R. Fried of the Pennsylvania Institute of Technology, I am writing to apply for a position as a geophysical engineer in your company.

I know that your company does work not only in Texas and Louisiana, but also in Saudi Arabia and Kuwait. I would be interested in a position that would enable me to work in foreign countries. You will notice from the attached résumé that I have a good command of French.

You will also notice that I have had some practical experience. May I point out that I worked in Greenland at the Alert Weather Station as one of a selected group. In addition to taking courses in geology and geophysics, I took a course in report writing; and I worked for one summer as an assistant to the editor of the Journal of Petrology.

I should be glad to come to Littleton for an interview, or perhaps you would like to arrange an interview through your company's personnel men who will be on our campus on March 2.

Sincerely,

Ralph J. Carrier

The second letter is accompanied by a fairly standardized kind of data sheet. However, no data sheet is completely standardized and at the same time successful, because the form must be adapted to each applicant, not the other way around.

Notice that the kind of job desired is given early. Most employers are impressed by well-trained students who know the kind of employment they want. The education is given in reverse order; the employer is likely to be more interested in the student's present than he is in the past; and he will be interested in the kinds of courses taken.

Companies are always looking for potential leaders. Therefore it is important that extracurricular activities and academic honors be given.

21.2.14 SPECIMEN JOB-APPLICATION LETTER (DATA SHEET, LETTER)

PERSONAL DATA

Ralph John Carrier
3505 Manfred Drive
College Park, Pennsylvania 16802

Permanent Address: 21 Summer Street
Rochester, New York 14601

Age: 26

Citizenship: United States

Marital Status: Married; one child

Physical Condition: Good

Military Record: Two years' active duty, U. S. Army

Rank in Class: Upper quarter of class.

Available: After June 15, 1970

EMPLOYMENT SOUGHT: Geophysical Engineer; prefer employment in oil company; prepared to work abroad.

EDUCATION:

College: Pennsylvania Institute of Technology
B.S. (Geological Sciences), 1970 (June)

Important Courses: Geology (50 credits); physics (15 credits); mathematics (15 credits); chemistry (15 credits); technical writing; effective speech; French (12 credits); anthropology.
Can read and speak French.

Secondary School: President McKinley High School, Rochester, New York. Graduated 1964.

EXPERIENCE:

Organization	Position	Dates
Journal of Petrology Pennsylvania Institute of Technology	Assistant to the Editor	July - August 1969
National Nursery Co. Rochester, N.Y.	Greenhouse work	July - August 1964 July - August 1967

COLLEGE ACTIVITIES:

Secretary, student chapter, National Geophysical Society
Features Editor, P.I.T. Engineer
Junior class representative, Student Council

The references cited should show the breadth of experience and not include friends and relatives unless they are pertinent to the student's professional background.

HONORS:

Junior and senior tuition scholarship
One of 25 physical science students selected for summer
work at Alert Weather Station, Greenland, 1968.

REFERENCES:

Dr. L. D. Sawyer, Chairman
Department of Geology
Pennsylvania Institute of Technology
College Park, Pa. 16802

Mr. F. R. Matthews, Director
Alert Weather Station
Greenland

Dr. G. R. Fried, Editor
Journal of Petrology
Pennsylvania Institute of Technology
College Park, Pa. 16802

21.2.15 Specimen Formal Report

The short report that follows is an example of a formal research report. As we have already pointed out, each organization and each person writing reports must adapt the techniques of reporting to suit individual needs. It would be impossible, as well as unrealistic, to assume that any single company report could serve as a model for all reports. This report issued by the National Aeronautics and Space Administration is, we think, a good example.

A standardized cover has been used. It includes title, names of authors, and file numbers.

NASA CONTRACTOR REPORT

NASA CR-1439

NASA CR-1439

DIFFERENTIAL THRESHOLDS FOR MOTION IN THE PERIPHERY

by James M. Link and Leroy L. Vallerie

Prepared by

DUNLAP AND ASSOCIATES, INC.
Darien, Conn.
for Electronics Research Center

NATIONAL AERONAUTICS AND SPACE ADMINISTRATION · WASHINGTON, D.C. · SEPTEMBER 1969

Following the cover is the title page. In addition to the items on the cover are a statement on distribution, a contract number, and the company's report number.

The decimal system of headings has been used. It first shows up in the table of contents, each item referring to a section in the text. In this report, the table of contents does not carry the list of tables and figures.

21.2.15 SPECIMEN FORMAL REPORT

NASA CR - 1439

DIFFERENTIAL THRESHOLDS FOR MOTION IN THE PERIPHERY

By James M. Link and Leroy L. Vallerie

Issued by Originator as Report No. BSD 69-701

Prepared under Contract No. NAS 12-88 by
DUNLAP AND ASSOCIATES, INC.
Darien, Conn.

for Electronics Research Center

NATIONAL AERONAUTICS AND SPACE ADMINISTRATION

ii

TABLE OF CONTENTS

LIST OF TABLES AND FIGURES

Figures | Page

Tables

The introduction orients the subject matter for the reader. It discusses the historical background and some of the theory. It then goes on to give the purpose of the study and what is expected of it.

The next section combines the summary and conclusions, showing in listed digest form how the problem was solved and what conclusions were derived.

Although the format of the body of this report has been derived from the material itself, the authors have kept to a fairly traditional outline. The method includes the procedure and is followed by the results and the discussion.

The recommendations conclude the main portion of the report. This section might be improved by some listing and numbering of important items.

The bibliography pages illustrate a variation in arrangement (see Sec. 17.5.2).

21.2.15 SPECIMEN FORMAL REPORT

DIFFERENTIAL THRESHOLDS FOR MOTION IN THE PERIPHERY

By James M. Link and Leroy L. Vallerie
Dunlap and Associates, Inc.

1. INTRODUCTION

Advances in our aerospacecraft capabilities have required man to perform an increasing number of difficult, continuous control tasks based on visual information presented to him on conventional displays designed for viewing with central vision. Typical examples of such tasks are landing high-performance aircraft during poor weather conditions and maneuvering or rendezvousing spacecraft using multidimensional control systems. The time required to move and refocus the eyes in visually switching between information sources seriously restricts the rate with which man can acquire information and hampers his control performance (Vallerie, 1967; Wulfeck, Weisz, and Raven, 1958; Travis, 1948). In an effort to eliminate the detrimental effects of visual switching, displays have been developed for viewing with peripheral vision. Majendie (1960), for example, proposed to use such displays to "provide flight intelligence to the pilot without distracting his attention from other tasks, without preventing him from looking freely about, either through the windscreen or within the cockpit, so that he can take appropriate corrective action from the information provided without serious interruption to his other tasks."

Both simulator and flight tests have indicated that valuable information can be obtained through peripheral vision "while" central vision is used to scan other information sources; e.g., looking for the runway and, at the same time, receiving control information by means of a peripheral display (Vallerie, 1967, 1968; Moss, 1964a, 1964b; Holden, 1964; Keston, Doxtades and Massa, 1964; Fenwick, 1963; Brown, Holmquist and Woodhouse, 1961; Chorley, 1961; Majendie, 1960). In this case, the operator is required to switch only his attention to information presented in his periphery instead of wasting precious time in redirecting and refocusing his eyes on spatially separated conventional displays.

Research carried out by Vallerie (1967), for example, clearly demonstrated that control performance deteriorates as visual switching increases and that peripheral displays can be used to overcome its adverse effects. The utility of peripheral displays, therefore, is well established, and a number of displays employing motion as the primary encoding stimulus have been developed to facilitate aircraft control especially during final approach and landing (Vallerie, 1968; Reede, 1965; Fenwick, 1963; Chorley, 1961). Displays incorporating other than motion as the primary encoding stimulus have not been given serious consideration because of their relatively limited discriminability in the visual periphery and their incompatability with anticipated operational environments. For example, displays requiring shape and pattern recognition, in addition to being relatively poor in the periphery, would suffer from accelerative forces and vibration; displays utilizing changes in color, brightness, or flicker rate would be affected by the level of ambient illumination in cockpits; while a velocity display involving motion would be satisfactory under a wide range of illumination. Research carried out by Salvatore (1968) also indicated that velocity information is more accurately assessed in the periphery than in the fovea.

2

Additional support for the use of motion as the primary stimulus for encoding peripheral displays is based in the physiological structure of the peripheral retina. A moving object, for example, causes the neural receptors to fire at their maximum excitation level continuously, while a stationary object causes the receptors to fire at their maximum level for only a brief period at the beginning of excitation. The level then decreases to a lower steady state. Neural receptors in the periphery are also connected in groups to a single synapse to produce summation of impulses. Hence, perception of moving objects is enhanced by both high excitation and summation of neural impulses, which helps to explain their attention-getting quality (Polyak, 1957; Granit, 1931; Adrian, 1928).

Few studies have been conducted dealing with the perception of motion in the periphery. Most of these studies have been concerned primarily with the definition of absolute thresholds of motion and the decreased sensitivity of peripheral vision (McColgin, 1960; Gordon, 1947; Klein, 1942). Apparently, no research has been conducted to determine differential thresholds for motion in the periphery, i.e., detection of changes in the rate of motion. McColgin, (1960), for example, mapped the periphery in terms of the absolute threshold of motion. He found that the threshold increases linearly as a funtion of eccentricity angle for both linear and rotary motion. Threshold isograms for both rotary and linear motion were elliptical in shape, with the horizontal axis approximately twice as long as the vertical axis.

The purpose of this study is to determine the differential thresholds for rotary and linear motion in the periphery. With definitive data on this visual ability, it will be possible to better assess the adequacy of motion cues for encoding information presented in the periphery, to determine the type and range of motion most suited for peripheral displays, and to select optimum locations for positioning them in the operator's visual field.

2. SUMMARY AND CONCLUSIONS

A study was conducted to determine the differential thresholds for rotary and linear motion in the periphery. Isograms were developed based on threshold estimates obtained from ten subjects using the psycho-physical method of limits. Based on the results of the study, the following conclusions can be made:

1. Differential thresholds for rotary and linear motion were found to increase as a linear function of eccentricity angle.

2. Threshold isograms for both types of motion are elliptical in shape, with the horizontal axis approximately twice as long as the vertical axis.

3. Based on statistical analysis, there appears to be no real difference between rotary and linear motion. Subjects, however, reported a preference for rotary motion.

4. Thresholds decreased logarithmically as a function of reference speed.

5. Age of the subject appears to be a highly significant factor influencing the perception of motion.

6. At high display velocities, subjects reported the occurrence of interference factors such as blur, fusion, strobing, and flicker. These factors were particularly noticeable with the linear display.

3. METHOD

Differential thresholds for linear and rotary motion were measured using ten subjects and the psycho-physical method of limits under controlled laboratory conditions. Three reference velocities were employed: 7.98 cm/sec, 19.95 cm/sec, and 31.92 cm/sec. Twenty-three different locations in the peripheral retina were investigated as illustrated in Figure 1. All possible combinations of motion, reference velocities, and retinal locations were presented to each subject in a different random order.

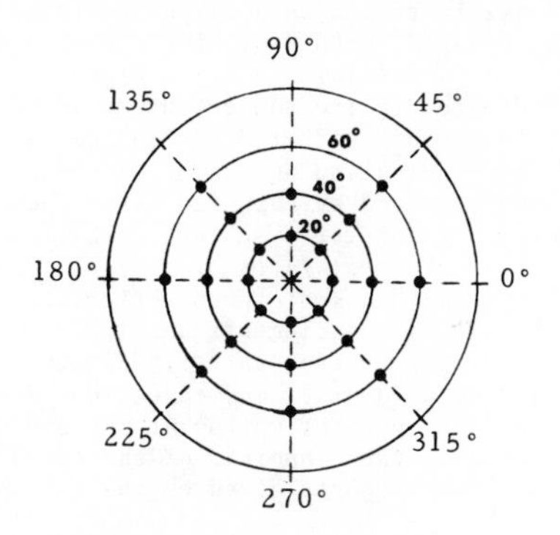

Figure 1. Location of Test Positions in the Visual Field.

3.1 Motion Displays

The two displays employed in the study were similar to standard aircraft instruments and were of the same aerial size. They were constructed of translucent lucite plastic and coated with flat black paint except for that portion of the display representing the moving element or "hand." The rotary display was 7.62 cm in diameter and contained a moving element which was 2.54 mm wide and 7.62 cm long. The hand rotated at its center in a counterclockwise direction. The linear display was 5.97 cm wide by 7.62 cm high, and it contained a hand identical to that used in the rotary display except that it moved from right to left in a linear fashion. The visual area subtended by the background and hand were identical for both displays. The displays were driven by a variable-speed motor. Display velocity was varied and controlled by means of preset potentiometers and a motor controller. Presentation time was maintained at 3 seconds by an electronic timer that automatically alternated between the preset reference velocity and the differential velocity or test speed. A tachometer generator and digital voltmeter were used to measure and monitor the speeds of the displays. They were situated 71.12 cm from the subject's eyes. A head-and-chin rest was used to stabilize the display image in the visual field.

3.2 Subjects

Ten subjects were selected from the staff of Dunlap and Associates, Inc. Selection was based on the results of visual tests carried out with a Ferre-Rand Perimeter and Keystone Telebinocular. All subjects who participated in the study possessed normal central and peripheral vision without the use of corrective lenses. Their ages ranged from 22 to 45 years.

3.3 Procedure

Each subject received standard instructions, which contained an explanation of the study's objectives, the method of response, and the necessity of maintaining the designated fixation point. Twenty minutes of practice was provided before the first session and a 5-minute warm-up prior to each subsequent session. A typical session lasted for a period of 30 minutes.

The method of limits was employed to measure the differential thresholds for each location in the visual field. Each reference velocity was presented for a period of 3 seconds and then immediately followed by a 3-second test period in which the speed was increased to a slightly higher rate. The reference velocity was then repeated immediately for a period of 3 seconds and followed again by a test period containing a still higher rate. During the test period, a tone was presented to the subject by means of earphones. The tone signaled the subject to respond either "yes" indicating that a change in speed was noticed, or "no" indicating no change was seen. The tone also served to mask auditory cues produced by changing motor speed. Speed was increased in stepwise increments until the subject verbally reported a difference between the reference and the test rates of motion. Threshold estimates were also obtained by starting with a test speed noticeably higher than the reference and descending in stepwise increments until the subject reported that both rates appeared identical to one another. Six threshold estimates were obtained: three in an ascending order and three in a descending order. Differential thresholds for both types of motion at any one retinal position were then calculated by averaging the midpoints of the speed intervals between the positive and negative reports given by the subjects.

4. RESULTS

Differential thresholds for both linear and rotary motion were calculated by averaging the threshold estimates for each reference velocity and test position in the periphery. In general, thresholds for both types of motion were found to increase linearly as a function of eccentricity angle along each radius of the visual field. Figure 2 shows the average threshold for all radii plotted as a function of eccentricity angle. Threshold values are expressed in the standard Weber form, $\Delta V/V$. Because of the linear increase in threshold with eccentricity angle, it was possible to interpolate between known threshold values along each radius in order to calculate points of equal threshold in the visual field. These data were then plotted in the form of isograms as shown in Figures 3 through 8. The general shape of the isograms, regardless of the reference velocity and type of motion, is elliptical, with the horizontal axis approximately twice as long as the vertical axis. This elliptical shape indicates that the subjects' ability to detect changes in motion was better on either side of this line of sight than above or below it.

Threshold isograms for the slowest reference velocity (Figures 3 and 4) indicate that there was little difference in performance with the two displays near the center of the visual field. However, from about 40 degrees outward, thresholds for rotary motion were lower than those for linear motion along the vertical axis. In contrast, the opposite was true at the medium and high reference speeds, especially along the horizontal axis, as illustrated in Figures 5 through 8. When thresholds were quite different across the entire visual field, thresholds for linear motion were generally lower than those for rotary motion. One exception occurred in the area near 20 degrees at the medium speed, where there was little difference between the two types of motion. As shown in the isograms for the high reference velocity, threshold values for rotary motion approached but never quite equaled those for linear motion as the eccentricity angle increased.

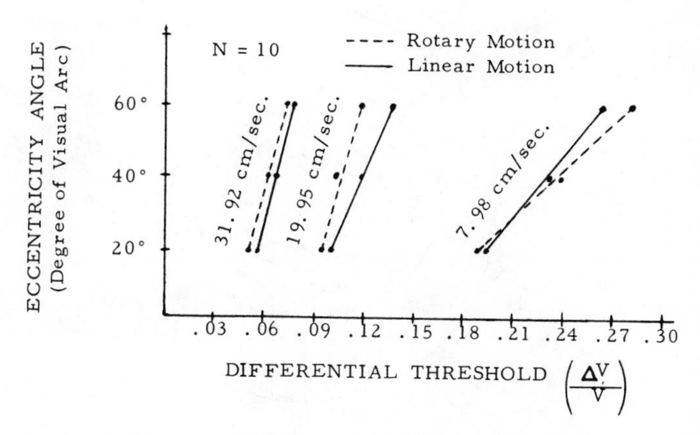

Figure 2. Differential Threshold as a Function of Eccentricity Angle.

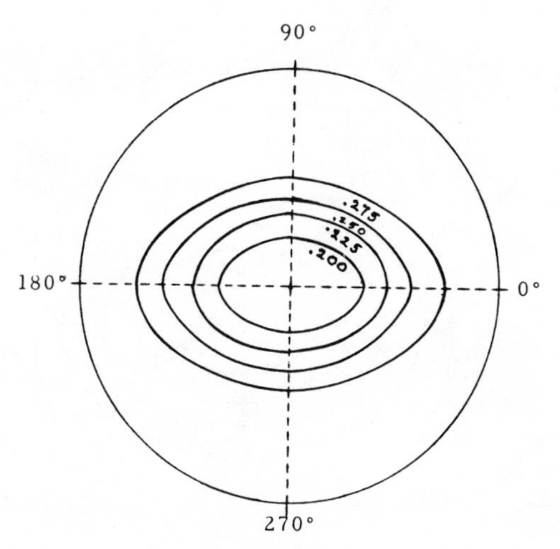

Figure 3. Differential Threshold $\left(\frac{\Delta V}{V}\right)$ Isograms for Linear Motion at the Slow Reference Velocity (7.98 cm/sec).

6

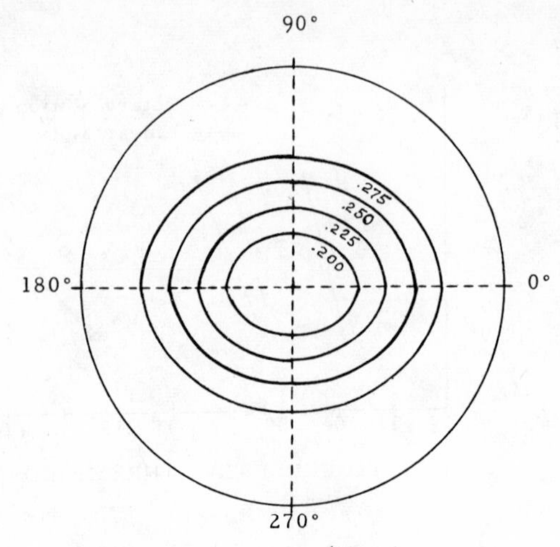

Figure 4. Differential Threshold $\left(\frac{\Delta V}{V}\right)$ Isograms for Rotary Motion at the Slow Reference Velocity (7.98 cm/sec).

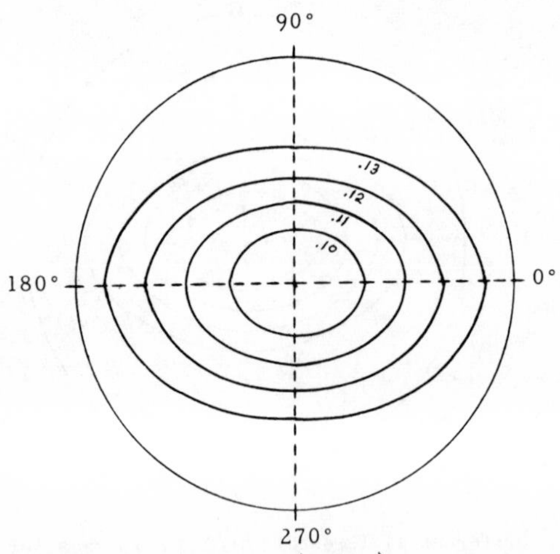

Figure 5. Differential Threshold $\left(\frac{\Delta V}{V}\right)$ Isograms for Linear Motion at the Medium Reference Velocity (19.95 cm/sec).

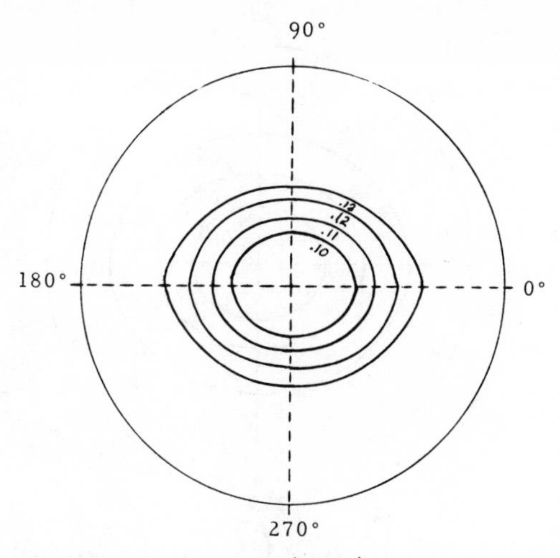

Figure 6. Differential Threshold $\left(\frac{\Delta V}{V}\right)$ Isograms for Rotary Motion at the Medium Reference Velocity (19.95 cm/sec).

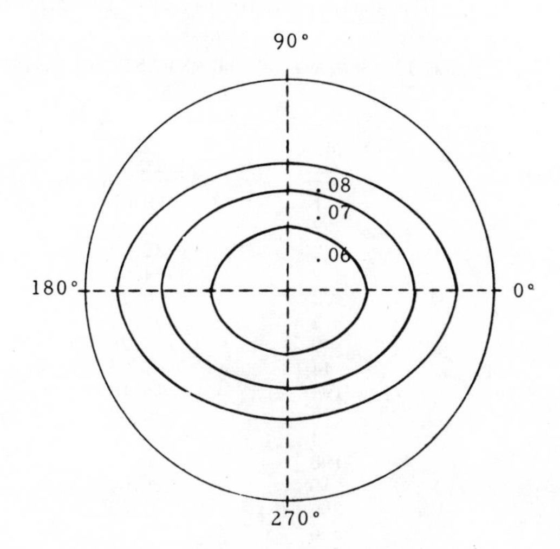

Figure 7. Differential Threshold $\left(\frac{\Delta V}{V}\right)$ Isograms for Linear Motion at the High Reference Velocity (31.92 cm/sec).

8

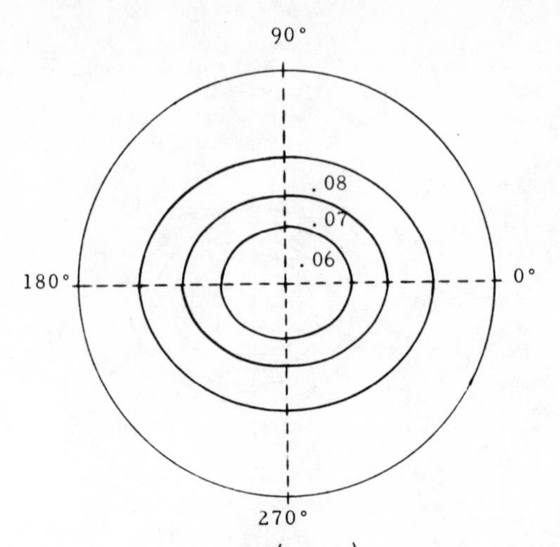

Figure 8. Differential Threshold $\left(\frac{\Delta V}{V}\right)$ Isograms for Rotary Motion at the High Reference Velocity (31.92 cm/sec).

An analysis of variance was conducted to test the differences between types of motion and reference velocities. The results of the analysis are contained in Table I. The analysis indicated that the difference between

TABLE I - SUMMARY OF THE ANALYSIS OF VARIANCE

Source	df	MS	F
Display	1	158.61	1.27
Position	22	197.33	N.A.
Velocity	2	732.90	4.18*
Subjects	9	1734.66	991.23*
D x P	22	8.74	1.08
D x V	2	105.79	0.32
D x S	9	125.24	
P x V	44	7.05	0.94
P x S	198	15.16	
V x S	18	175.20	
D x P x V	44	6.71	0.80
D x P x S	198	8.11	
D x V x S	18	326.64	
P x V x S	396	7.53	
D x P x V S	396	8.44	
Within	6900	1.75	
TOTAL	8279		

*Significant at the 0.05 level or better.

the two types of motion was not significant and could be attributed to chance fluctuation of the threshold estimates. Only two factors, reference velocity and subjects, were found to be significant. Scheffe's test revealed that only the difference between the slow and medium speeds could be accepted as real; it accounts for the significance of the speed factor in the variance analysis. Dividing the subjects into two age groups, 21 to 30 and 31 to 45 years, average thresholds were found to be 1.85 cm/sec and 2.29 cm/sec, respectively. The difference between these means was highly significant. Age, therefore, appears to have an adverse effect on the perception of motion, as is the case with many other visual functions. This finding, therefore, was not completely unexpected. Further research, however, will be required to investigate this effect more thoroughly.

The effect of speed on differential threshold was of particular interest in this study. In Figure 9, threshold is plotted as a function of reference velocity. In accordance with Fechner's "law," the differential threshold ($\Delta V/V$) decreased, in a logarithmic fashion, as a function of reference speed regardless of the eccentricity angle and the motion type; i.e., the percent change in speed required for detection decreased at higher display speeds. This relationship between threshold and reference speed was hypothesized and not unexpected since other psycho-physical functions also assume this same general form.

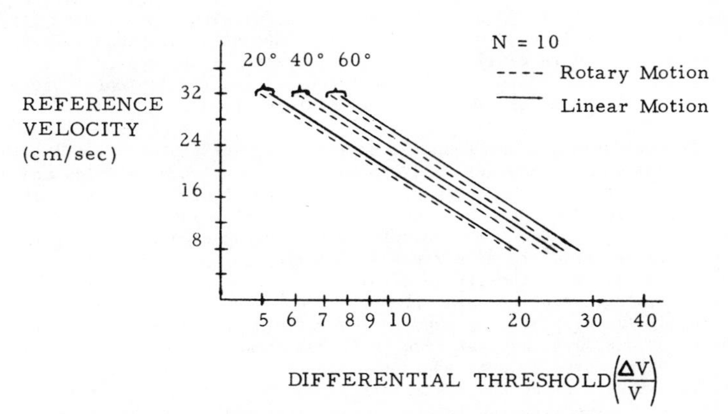

Figure 9. Differential Threshold as a Function of Reference Speed for Three Eccentricty Angles.

5. DISCUSSION

The results of the study clearly demonstrated that the ability to differentiate changes in the rate of motion decreases from near the fovea to the outermost areas of the periphery. Isograms for both rotary and linear motion were found to be elliptical in shape, with their horizontal axis extending approximately twice as far as the vertical axis. Similar isograms were reported by McColgin (1960) for the absolute thresholds of motion in the periphery. These findings might be explained by several hypotheses and established facts regarding the physiological structure of the retina. Polyok (1957) and Granit (1930), for example, suggest that two contributing factors are the decrease in the number of receptors in the periphery and the summation of impulses resulting from the synaptic convergence of neurons of the peripheral receptors. In addition, Polyok (1957) also indicated that the decrease in sensitivity from the fovea to the periphery may result from differences in the responsiveness of cones located in different areas of the retina. Wolf (1959) hypothesized that the elliptical shapes of the threshold isograms may be caused by the pattern of retinal innervation in relation to the blind spot. While these hypotheses suggest several explanations for the present findings, the lack of additional physiological data limits their acceptability. Additional research, therefore, is required before the response characteristics of the peripheral field can be adequately explained.

A comparison of the differential thresholds for rotary and linear displays was considered feasible since the visual area subtended by both displays was equal. Furthermore, the reference and test velocities employed in the study were the same for both displays. With the rotary display, tip speed was suggested by McColgin (1960) as being the most significant determinant in the perception of rotary motion. Additional support for this contention was provided by the reports of the subjects in this study. Hence, tip speed of the rotary display was equated with the speed of the linear display and the two compared quantitatively. Although there appeared to be a difference between the two displays based on the threshold isograms, a statistical comparison indicated that this difference was not significant. Introspective reports of the subjects, however, indicated a preference for rotary rather than linear motion. Typically, subjects said that rotary motion was ". . . more easily seen," ". . . easier on the eyes," ". . . more confortable to view." Similar statements concerning the subjects' preference for rotary motion were recorded by McColgin (1960).

Reference velocity was found to be a significant factor affecting the threshold estimates. However, only the difference between the slow and medium velocities could be accepted as real. The absolute threshold at the high velocity fell between those obtained at the slow and medium velocities. A confounding factor that may account for this unexpected result was that all subjects reported the occurrence of interference phenomena at the high speed, especially with the linear display. These phenomena consisted of blurring, fusing, strobing, and flicker. Two subjects also indicated that occasionally two displays appeared in their field of view simultaneously. Of these two displays, one was seen to travel at an extremely high rate. while the other moved in a stroboscopic manner. One subject reported blur and fusion while viewing the linear display at the medium reference speed. Blurring and fusing with the rotary display at the high velocity were also reported by a few subjects. The occurrence of these phenomena is not understood. Blurring may be explained in terms of flicker fusion. However, further research is required to develop specific explanatory hypotheses and to study them under controlled conditions in the laboratory. It appears appropriate to state, however, that high velocities should be avoided in the design of peripheral vision displays.

6. DESIGN IMPLICATIONS AND RECOMMENDATIONS

Based on the results of this study, it is possible to make specific recommendations with regard to the design and location of peripheral displays in the visual field. It was demonstrated that the ability to discriminate changes in the rate of motion is relatively good in the periphery, even out to 60 degrees in the visual field. Motion, therefore, appears to be a suitable stimulus dimension for encoding peripheral displays.

No significant difference was found between linear and rotary motion. Although the subjects who participated in the study stated a preference for rotary motion, in view of their performance data, either type of motion appears to be adequate for providing control information to vehicle operators. The type of motion employed in peripheral displays should be based on the consideration of the operational situation and the type of control information it will present to the operator. For example, there appears to be no simple way, utilizing a rotary display, for presenting directional information such as aircraft pitch and roll. Linear displays, such as the Collins PCI and the Smith PVD, have been successfully employed for this purpose.

At high velocities, the subjects witnessed the occurrence of certain interference phenomena such as blurring, strobing, fusing, and flicker, especially with the linear display. These phenomena cannot be easily explained without further investigation. However, they do suggest that display velocity should be limited to some value below 30 cm/sec if these phenomena are to be avoided in the operational situation. Although an absolute upper limit cannot be determined on the basis of this study, velocity should probably be limited to approximately 20 cm/sec, the medium reference velocity investigated. Only one subject reported blurring and fusing at this velocity with the linear display.

The results of the study also indicate that performance was approximately twice as good along the horizontal as compared to that along the vertical meridian, i.e., isograms of differential motion thresholds were elliptical in shape, approximately twice as wide along the horizontal axis. Since performance degrades at a slower rate along the horizontal axis, displays may be located at greater angular distances from the normal line of sight for the same level of performance. Therefore, peripheral displays are best located on either side of the operator and as near level as possible with his normal line of sight. The angular distance, of course, should be kept to a minimum in order to assure the highest degree of performance.

12

7. BIBLIOGRAPHY

Adrian, E. D. The basis of sensation, London: Christophers, 1928.

Brown, I., Holmquist, S., and Woodhouse, M., A laboratory comparison of tracking with four flight director displays. Ergonomics, 1961, 4, 229-251.

Chorley, R. A. The development of the Smith Para-Visual Director: A lecture to be presented to the Bristol branch of the Royal Aeronautical Society on Thursday, 26 October, 1961. R. I. D. 597, Smith Industries, Ltd., England, October 1961.

Fenwick, C. Development of a peripheral vision command indicator for instrument flight. Human Factors, 1963, 5, 117-128.

Gordon, D. A. The relation between the thresholds of form, motion, and displacement in parafoveal and peripheral vision at a scotopic level of illumination. Amer. J. Psychol. 1947, 60, 202-225.

Granit, R. Comparative studies on the peripheral and central retina, I. On interaction between distant areas in the human eye. Amer. J. Physiol., 1930, 94, 41-50.

Holden, K. J. Instrument displays for blind flying. J. Royal Aero. Soc., 1964, 68, 833.

Keston, R., Doxtodes, D., and Massa, R. Visual experiments related to night carrier landing. Human Factors, 1964, 6, 465-473.

Klein, G. S. The relation between motion and form acuity in parafoveal and peripheral vision and related phenomena. Arch. Psychol., 1942, 39, (No. 275), 1-70

McColgin, F. H. Movement thresholds in peripheral vision. J. Opt. Soc. Amer., 1960 50, 774-779.

Majendie, A. M. A. Para-Visual Director. J. Inst. Nav. (Brit), 1960, 13, 447-454.

Moss, S. M. Tracking with a differential brightness display: I. Acquisition and transfer. J. Appl. Psychol., 1964a, 48, 115-122.

Moss, S. M. Tracking with a differential brightness display: II. Peripheral tracking. J. Appl. Psychol., 1964b, 48, 249-254

Polyak, S. The vertebrate visual system. Chicago: University of Chicago Press, 1957.

Reede, C. H. KLM - Research on the lowering of weather minima for landing of aircraft. De Ingenieur, 77, No. 11, LI-13 (1965). Royal Aircraft Establishment, Library Translation No. 1250.

Salvatore, S. The estimation of vehicular velocity as a function of visual stimulation. Human Factors, 1968, 10, 27-32.

BIBLIOGRAPHY (cont.)

Travis, R. C. Measurement of accommodation and convergence time as part of a complex visual adjustment. J. Exp. Psychol., 1948, 38, 395-402.

Vallerie, L. L. Displays for seeing without looking. Human Factors, 1966, 8, 507-513.

Vallerie, L. L. Peripheral vision displays. NASA: CR-808, 1967.

Vallerie, L. L. Peripheral vision displays: Phase II Report, NASA: CR - 1239, 1968.

Vallerie, L. L., and Link, J. M. Visual detection probability of "sonar" targets as a function of retinal position and brightness contrast. Human Factors, 1968, 10, 403-412.

Wolf, E. (Personal communication) 1959. In McColgin, F. H. Movement thresholds in peripheral vision. J. Opt. Soc. Amer., 1960, 50, 774-779.

Wulfeck, J. W., Weisz, A., and Raben, M. W. Vision in military aviation. WADC Tech. Rept. 58-399, Wright-Paterson Air Force Base, Ohio. 1958.

21.2.16 Specimen Formal Report (Sample Pages)

The report that follows was written by a graduate student as one of the requirements for the degree of Master of Engineering. Because the original was so long, only representative portions have been included here.

As is customary in a formal report, the items on the title page are presented in as symmetrical a manner as possible. In a student report such as this one, certain items that might be found in a company report are missing.

21.2.16 SPECIMEN FORMAL REPORT (SAMPLE PAGES)

DENSITY - STRUCTURE RELATIONSHIPS

IN PYROLYTIC GRAPHITE

By

Franz J. Buschmann

Submitted to the Faculty of the

Materials Division of the School of Engineering

Rensselaer Polytechnic Institute
Troy, New York

January, 1969

The table of contents provides a comprehensive outline of the report down to the third degree. The numeral-letter system used offers a contrast with the outline system of the formal report of Sec. 21.2.15.

21.2.16 SPECIMEN FORMAL REPORT (SAMPLE PAGES)

iii

TABLE OF CONTENTS

In this report, the list of tables is short. It might well have been added to the list of figures.

LIST OF TABLES

21.2 SPECIMENS

LIST OF FIGURES

Most authors of reports find that they must depend to some extent on the help of other people. The section on acknowledgments does not attempt to go into detail on the help received. But it does express appreciation for assistance that often cannot be measured in specific terms.

ACKNOWLEDGMENTS

The author would like to thank Dr. R. J. Diefendorf, not only for his constant help and guidance throughout this work, but also for supplying the samples that were used.

Sincere thanks are due to B. Butler and R. Mehalso for their helpful comments and discussions, with a special note of thanks to B. Butler for his assistance in obtaining the electron microscopy and x-ray diffraction data.

The abstract of this report is divided into three parts: a brief statement on what had already been done on the subject of the report, the reason for extending the investigation, and the results obtained.

As with many research reports, this one cannot be as cut and dried in the results section as other types of engineering reports. This particular abstract seems to lie between the purely descriptive and the purely informative.

ABSTRACT

The properties and structure of pyrolytic graphite have been reported by many investigators. The data often seem inconsistent and contradictory. The reasons for this apparent inconsistency include incomplete specifications of all the deposition variables, so that exact duplication is impossible, and the often-incorrect assumption that the structure of pyrolytic graphite is an invariant.

The purpose of this investigation was to examine the structure of pyrolytic graphite and relate it to the deposition conditions and observed properties. Density and stress measurements were performed to get an indication of where structure changes could be expected. Optical and electron microscopy and x-ray diffraction were employed to determine the structures.

The results showed that a pyrolytic graphite deposit can consist of a number of structures, especially if deposited around 1700 and 1900 C, and that the types and amounts of the structures are dependent on the deposition conditions. While the density of low-temperature and high-temperature deposits were dissimilar, the degree of preferred orientation was nearly identical, indicating that perhaps two different types of pyrolytic graphite were being compared.

This introduction is an important part of the report. It orients the reader to the conditions leading to the present investigation and is particularly helpful in pointing out some of the problems involved.

The last paragraph describes the purpose of the study together with a brief statement on the procedure.

21.2.16 SPECIMEN FORMAL REPORT (SAMPLE PAGES)

I. INTRODUCTION

Pyrolytic graphite has been produced and may be produced by passing a hydrocarbon gas over a suitable substrate at an elevated temperature and usually at a low pressure. The substrate is generally a commercial graphite rod or cylinder but may be a ceramic rod or plate, silica spherules,[1,2] or nuclear-fuel particles. The gas is usually methane or benzene, with or without a diluent gas. But ethane, propane, acetylene, or other hydrocarbons may also be used. A low pressure—several torr to tens of torr—is usually maintained in the deposition vessel, though pyrolytic graphite can be deposited at atmospheric pressure, using a few-percent hydrocarbon gas in a suitable carrier gas.[3] The deposition temperature may vary between 900 and 2600 C, depending on the specific application or structure desired. Usually, pyrolytic graphite for structural applications is made at temperatures between 2000 C and 2200 C.

A description of the decomposition of pyrolytic graphite is complicated because of the large number of competing processes. The processes include gas-phase reactions, diffusion to and from the surface, and surface reactions. These processes and their relative importance are dependent on a large number of variables, including temperature, pressure, gas velocity, initial gas composition, and furnace geometry. The deposit and the rate of deposition, then, depend on the specific operating conditions and the resulting importance of the competing reactions. Most of the data on pyrolytic graphite do not include a complete statement of the deposition conditions; usually only the temperature is specified. But the structure, and consequently the properties, can be highly dependent on all the deposition conditions.

The purpose of this investigation is to relate the deposition conditions to structure by making a thorough study of pyrolytic graphite deposited under varying conditions. Studies have been made on the change in the deposition rate and of the structure of pyrolitic graphite caused by varying temperature, pressure, gas flow, position in the furnace, and—to a lesser degree—composition of the feed gas on the deposition rate and structure of pyrolitic graphite. Optical microscopy, electron fractography, and x-ray diffraction were employed to determine the structure. Since density is sensitive to structure, density and stress measurements were performed on a large number of samples deposited under different conditions.

The author of this report felt it necessary to include a fairly complex historical review. Such a review is found in most reports dealing with comparatively new research fields. The review, as in this case, will serve to educate the reader as well as to inform him.

II. HISTORICAL REVIEW

A. The Deposition Process

The fact that carbon may be produced by passing a hydrocarbon gas over a heated surface has been known for centuries. But only until fairly recently has there been much scientific investigation of vapor-deposited carbon. Brown and Watt[4] conducted some preliminary studies on pyrolytic graphite deposited from methane onto a small graphite rod, resistance-heated to between 1600 and 2600 C, at a pressure of about 15 torr. They reported some density measurements and performed some x-ray diffraction and optical microsocpy studies.

The apparatus for depositing pyrolytic graphite has been described by many authors.[5,6,7] Figure 1 shows the apparatus of Diefendorf and Butler.[8] Other furnaces may have modifications, but they consist basically of a vacuum chamber, a heat source (in this case a resistance-heated tube of pyrolytic graphite), a port for introducing gases, and a means of maintaining a vacuum. The apparatus depicted in Figure 1 can be described as a hot-wall furnace.

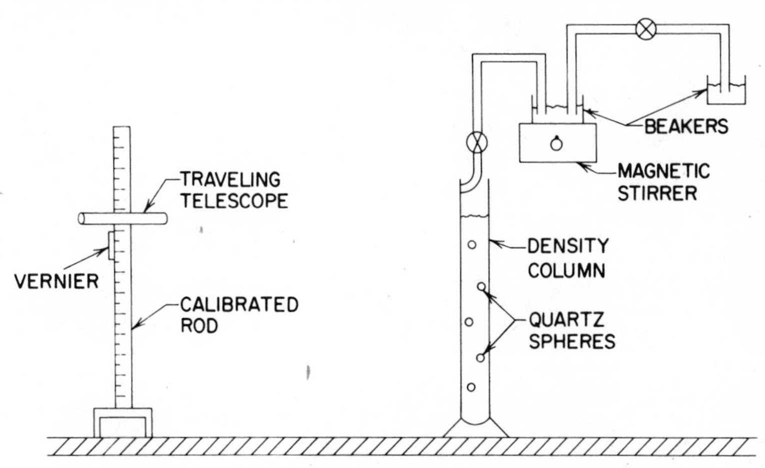

Figure 1 Schematic Drawing of the Deposition Furnace

Another basic type of apparatus is a cold-wall furnace,[5,7] where the heating element is not the furnace wall, but a central element. The central element could be a resistance-heated filament or a small carbon rod. In this type of furnace the deposition temperature is dependent on the resistance of the element, which depends on the cross-sectional area. Since pyrolytic graphite is depositing on the heater element during the run, the area is changing, and the temperature will vary. Also, there is the problem of uneven deposition and consequently uneven temperatures. In addition, there is the problem of a wide range of gas temperature. The gas temperature varies from the deposition temperature at the mandrel to almost room temperature near the furnace walls. However, although the description of the deposition process is more complicated than in hot-wall furnaces, most of the data reported in the literature for pyrolytic graphite is for the cold-wall type of furnace.

The above discussion is for pyrolytic graphite in general. It is possible to separate the field into two parts—pyrolytic graphite deposited on a stationary mandrel, and pyrolytic graphite deposited in a fluidized bed. This discussion will be limited to pyrolytic graphite deposited on a stationary mandrel.

21.2 SPECIMENS

The variables for pyrolytic graphite deposited in a fluidized bed are radically different from those encountered during deposition on a mandrel. Since the deposition surface area is very large, the concentration and flow rate of the gas is important. In addition, the gas phase reactions are diminished because of the relatively small diffusion distances between particles. For these reasons and others, the resulting strucutres and properties, and how they are attained, are significantly different from those found in pyrolytic graphite deposited on a stationary mandrel. For a complete review of the deposition of pyrolytic graphite in a fluidized bed, J. C. Bokros[3,9] gives a detailed account.

B. The Deposition Mechanism of Pyrolytic Graphite

The deposition mechanism of pyrolytic graphite is difficult to describe because of its complexity. Any one of a number of steps may be limiting, or of primary importance at any given set of conditions.

Qualitative models and theories have been proposed. Bokros[9] reviews some of these, in particular the explanation offered by Tesner[10] and others similar to it. Tesner sees the pryolysis as a competition between two processes—a competition between the homogeneous nucleation of gas-phase particles and the direct condensation of carbon-bearing specie on a heated surface.

Others[11,12] have attempted to explain the deposition. . . .

The materials and apparatus were basic to this research project. Therefore, a considerable amount of space is devoted to these two items.

The author of this report has described the source of the materials, and he has reinforced his description of the apparatus by providing drawings that are incorporated into the text. He has used headings effectively to set off the individual items in the overall section on apparatus.

III. MATERIALS AND APPARATUS

A. Materials

The various types of pyrolytic graphite used in this thesis were produced at the General Electric Research Laboratory, Schenectady, New York. The graphite was made in an apparatus similar to that shown in Figure 1.

The samples consisted of pyrolytic graphite deposited inside an 8-inch-long tube of commercial graphite with an inside diameter of 1/2 inch and an outside diameter of 1-1/4 inches. This 8-inch tube was cut, perpendiuclar to the tube axis, into 3/4-inch samples. These were the samples as they were received from the General Electric Research Laboratory.

The pyrolytic graphite was deposited under different conditions. The samples under investigation were deposited from methane at temperatures varying from 1500 to 2300 C, pressures ranging from 2.5 to 50 torr, and flow rates between 0.11 and 4.0 cubic feet per hour. These were the primary variables under investigation. Others which were briefly studied included different gas compositions, particularly acetylene-hydrogen mixtures, different diameter deposition tubes, and different deposition tube thicknesses.

B. Apparatus

1. Density Measurements

Figure 7 is a schematic drawing of the apparatus used to measure the density. The density measurements were made by noting the position of the material in a density column. The density column consisted of a liquid of continually varying density. This liquid was prepared by carefully mixing two liquids.

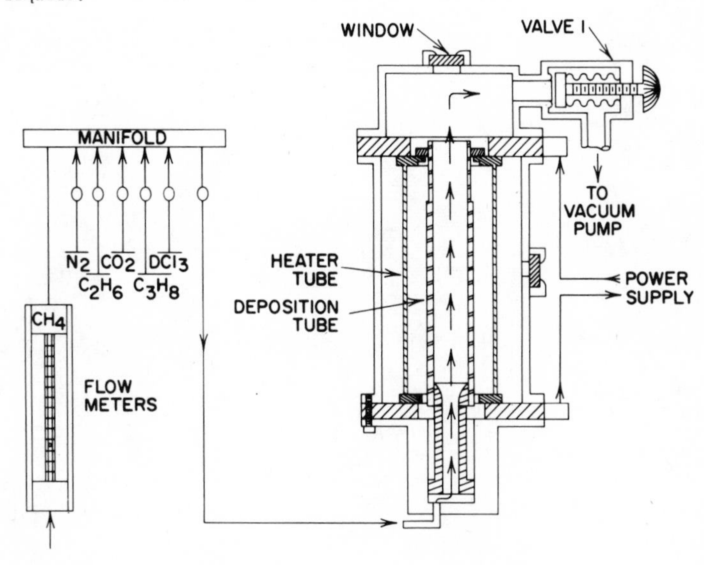

Figure 7 Schematic Drawing of the Density Apparatus

21.2 SPECIMENS

The column was prepared by continually mixing a flow of low-density liquid with a fixed amount of high-density liquid and continuously siphoning off the mixed liquid into the column. A column with a uniformly varying density will result if this is done properly. The density of the column will vary from that of the high-density starting liquid to that of the low-density liquid.

To cover the entire range of densities encountered, two different columns were prepared. They differed in the starting liquids used. A high-density column was prepared by mixing bromoform (from Eastman Organic Chemicals, No. P45) with high-purity carbon tetrachloride. The low-density column used a mixture of bromoform and carbon tetrachloride in the ratio of 1:2 or 1:3 and high-purity benzene as the low-density liquid.

To calibrate the column, a set of quartz spheres with known densities were placed in the column. The density of these spheres was known to ± 0.0005, and varied from 1.2000 to 2.6000 g/cc in intervals of 0.1000 g/cc.

To read the position of the material and the quartz spheres in the column, an Eberbach cathatometer was used, as indicated in Figure 7.

2. Electron Microscopy

Replicas for electron microscopy were prepared as described in Part IV. These replicas were examined in a 50-kv Hitachi 7-S electron microscope at 10,000 and 20,000 x.

3. Optical Microscopy

Metallographic samples were examined under polarized light, using a Bausch and Lomb Metallograph No. XR4306. To obtain the proper contrast, Kodak Plus-X Pan film was used.

4. X-Ray Diffraction

A Phillips-Norelco X-Ray Diffractometer, Type No. 12045B/3, was used to obtain the x-ray characteristics of the deposits.

To obtain the preferred orientation measurements, a special holder was used. This holder, designed and built by Barry Butler of Rensselaer Polytechnic Institute, consisted of a special stage inside a bearing holder. The stage was attached to a pulley, which was driven by a small motor. With the use of this holder, it was possible to rotate the specimen at a constant rate in a plane perpendicular to the x-ray beam. . . .

This section of the report combines results with a discussion.

An interesting feature in this report is that there is a continual reference to figures and tables. This seems to be necessary because the investigation depended so much on complicated data needing graphic explanation.

The author uses this section to point out inconclusive results, observations, and possible trends.

V. RESULTS

The results of the stress measurements are shown in Figures 8 and 9. These figures are plots of the residual stress, σ_t, versus t/R, the thickness-to-radius ratio. This is the usual method of comparing residual stresses of pyrolytic graphite. The cool-down stress, calculated from average thermal expansion coefficients, is also shown on these graphs. Appendix I gives the calculation for this cool-down stress. In Figure 8, the stresses for samples of a particular tube are connected to indicate the trends within a particular run. The numbers next to each stress value in Figure 8 give the distances, in inches, from the tube inlet for that sample.

Some of the density data are given in Figures 10 to 13, which are plots of density versus position in the tube. Figure 10 is a constant-flow graph, and Figure 11 is a constant-pressure graph for 1800 C deposits. Figure 12 consists of density-versus-position plots for different temperatures, while Figure 13 compares 1500 C deposits. These graphs are given to indicate the type of plots obtained from the density measurements. It should be noted that several curves were shifted laterally, right or left, where evidence indicated that experimental differences between runs had produced these shifts originally.

The remaining density plots are broken down into three values which are reported in Table I. The values were obtained by idealizing the plots shown in Figures 10 to 13, and represent approximate values only. Figure 14 shows how the plots were broken down and indicates the meaning of the reported numbers.

Figure 15 consists of three electron micrographs of replicas of the fracture surfaces of three different samples. The samples were deposited at 1500, 1700, and 1900 C, and the micrographs indicate the changes in the structure due to the increase in temperature. An attempt was also made to correlate changes in the observed microstructure with other deposition variables, but the results were inconclusive.

The x-ray diffraction data are given in Table II. The positions of the three major peaks are given, the half-angle value, $\phi_{1/2}$, is given, and a value of L_c calculated from the 002 peak broadening is reported. From independent calculations the machine broadening was found to be 7.75° in the ϕ direction and 0.25° in the θ direction. The values reported in Table II represent corrected values. Appendix II shows the calculation for the crystallite size, L_c. Figure 16 is an example of the preferred-orientation data.

The result of the optical microscopy was a series of 22 photographs showing the structure of a pyrolytic graphite deposit down an entire tube. Figure 17 shows six of these which illustrate all the significant structures observed. The distance from the inlet for each micrograph is also given.

Figure 17(a) shows the low-temperature structure near the inlet of the tube. Large featureless cones are typical of this structure. Figure 17(b) is further up the tube showing a normal, surface-nucleated pyrolytic graphite deposit, with just the beginnings of some continuous nucleation visible near the top of the deposit. Figure 17(c) shows the continuous nucleated structure. This structure was observed in a narrow band approximately 1 inch from the beginning of the tube.

In the next 4 inches, a normal, surface-nucleated deposit was observed. It consisted of fairly large cones with a fine fibrous texture. Figure 17(d) is characteristic of this structure. Small arc-shaped delaminations were visible in some of the cones. Figure 17(e) is near the end of the tube where a possible drop in temperature produced a supersaturation, and some continuous nucleation is evident. Figure 17(f) is the typical low-temperature structure, similar to that observed in Figure 17(a). . . .

Compare the summary with the abstract on page 335. The two elements show similarities, but the summary has been modified in the light of preceding material with which the reader has become familiar.

VII. SUMMARY

The density, stress, and preferred-orientation measurements, and the structure observations, have led to several important conclusions concerning pyrolytic graphite deposition. The properties, including stress and preferred orientation, are quite insensitive to deposition variables, unless the change in deposition variables produces a different structure.

The deposit depends on the type of specie from which deposition occurs. At high temperatures (greater than 2000 C), deposition is occurring almost exclusively from a small radical. At low temperatures (1500 C and at the beginnings of the tube at 1700 and 1900 C), deposition is occurring from a larger specie containing more hydrogen. At intermediate temperatures, a combination of the two types of pyrolitic graphite is found.

From preferred orientation and optical microscopy it is seen that the low-temperature structure is quite similar to the high-temperature structure, while the density is much lower. This suggests that either the reduced crystallite size is causing this drop in density, or that deposition is occurring from two different types of specie.

21.2 SPECIMENS

VIII. BIBLIOGRAPHY

1. Grisdale, R. O., "The Formation of Black Carbon," J. Appl. Phys., 24, 1082 (1953).

2. Grisdale, R. O., Pfister, A. C., and van Rossbroeck, W., "Pyrolytic Film Resistors: Carbon and Borocarbon," Bell Systems Tech. J., 30, 271 (1951).

3. Bokros, J. C., "The Structure of Pyrolytic Carbon Deposited in a Fluidized Bed," General Atomic Report No. GA-5163 (1964).

4. Brown, A. G. R., Hall, A. R., and Watt, W., "Density of Deposited Carbon," Nature, 172, 1145 (1953).

5. Diefendorf, R. J., "The Deposition of Pyrolytic Graphite," J. Chim. Physique (1960), pp. 815-821.

6. Brown, A. R. G., and Watt, W., "The Preparation and Properties of High-Temperature Pyrolytic Carbon," Ind. Carbon and Graphite, Soc. of Chem. Ind. (1958), pp. 86-100.

7. David, C., Sublet, P., Auriol, A., and Rappeneau, J., "Properties Structurales des Carbones Pyrolytiques Deposes Entre 1100 et 1800 C.," Carbon, Vol. 2, pp. 139-148 (1964).

8. Butler, B. L., "The Effect of Impurities on the Deposition of Pyrolytic Graphite," Master's Thesis, Rensselaer Polytechnic Institute (August, 1967).

9. Bokros, J. C., "The Deposition, Structure, and Properties of Pyrolytic Carbon," Gulf General Atomic Report No. GA-8300 (1967).

10. Tesner, P. A., Symp. Combust. 7th, Oxford, London, p. 546. (1959).

21.2.17 Specimen Formal Report

This report is typical of the short formal report prepared by a consulting firm for a client.

The cover would be made of cardboard, with the name of the issuing organization printed at the top. The title of the report is typed on a label in the box near the center.

The cover would be in the form of a folder, with a back leaf that is usually plain on both sides. The back leaf of the cover has been omitted from this specimen.

[Cover]

ARGYLE TESTING LABORATORY

Milray, California

EXAMINATION OF A HIGH-SPEED STEEL
CUTTER INSERT

for
Albermarle Tool Company

February 24, 1971

This title page repeats all the information on the cover, and adds the name of the writer and the addresses of the issuing and receiving organizations.

Notice that the logical blocks of information are represented by typographical blocks.

21.2.17 SPECIMEN FORMAL REPORT

EXAMINATION OF A HIGH-SPEED STEEL
CUTTER INSERT

for

Albermarle Tool Company
Reno, Nevada

Report by

Joseph J. Connelly, Head
Metallurgy Department

Argyle Testing Laboratory
1624 North Wexter St.
Milray, California

February 24, 1971

The table of contents reproduces the heads and subheads from the text.
The list of figures is sometimes put in a separate table; often it is omitted entirely.

CONTENTS

This foreword provides general orientation and describes the purpose of the investigation. It tells who authorized the work and states that recommendations were requested.

FOREWORD

This report covers the investigation of a high-speed steel cutter insert submitted to the Laboratory by the Albermarle Tool Company on February 6, 1971. The insert was from a batch of 25 all of which were heat treated at the same time. All 25 inserts were found to be cracked after the heat treatment.

Mr. H. L. Quentin, of the Albermarle Tool Company, asked Argyle Testing Laboratory to:

(1) Determine the cause of the crack.

(2) Ascertain whether the Dunster salt bath used for the heat treatment could have been a primary factor in the failure.

(3) Recommend steps to prevent repetitions of the failure.

This summary follows the foreword. It depends on the foreword for background information. After telling very briefly what tests were performed, it gives most of its attention to the conclusions and recommendations. It presents all of the most important information contained in the report.

The recommendations can sometimes be integrated with the conclusions in a single subsection. In this report the recommendations are in a subsection separate from the conclusions because half of the conclusions do not lead to recommendations.

SUMMARY

The cracked cutter insert was subjected to a metallographic examination and an abbreviated chemical analysis. These investigations lead to the following conclusions and recommendations:

Conclusions

(1) The crack existed in the insert before the heat treatment.

(2) The crack was caused by a flaw in the steel from which the insert was made.

(3) The hardened insert had a soft case that contained austenite.

(4) The surface of the insert was contaminated with nickel from the electrode of the Dunster salt bath.

(5) The surface also contained excess carbon and nitrogen, indicating that the salts used in the bath were not neutral.

(6) Because the carbon and nitrogen, as well as the nickel, could have have been factors in stabilizing the austenite and causing the soft case, the exact role of the nickel cannot be determined.

Recommendations

(1) The cracking of the cutter inserts should be taken up with the supplier of the steel from which they were made.

(2) Frequent pH determinations of the salt bath should be made, and the salts kept neutral during the heat-treating operation.

The discussion in this short report depends on the foreword for background information. After providing a little further history of the specimens, it starts to give results and conclusions mixed right in with descriptions of the work done.

Following the statements of several results and conclusions on page 5, the subsection on "Chemical Analysis" describes some of the test work done. Then follow more discussion and conclusions. The integration of results, discussion, conclusions, and description of method into one section produces a logical, clear account. Separate sections on methods and results in a report like this one would result in compartmentalization and repetition.

DISCUSSION

The cutter inserts were heat treated in a Dunster salt bath, which uses nickel and mild-steel immersed electrodes. It was suspected that this combination had contaminated the pieces with nickel, and that the contamination had been a factor in the cracking.

The crack appeared in the middle of the back face of the insert. This is the least likely place for cracking to occur during hardening, because the back face is flat, while the front face has many sharp corners and notches.

The cracked insert was subjected to a metallographic examination and an abbreviated chemical analysis.

Metallographic Examination

The crack was very straight, extending over nearly the whole length of the part and completely bypassing two tapped holes, the only irregularities on this surface. When the insert was sectioned, it was discovered that the crack was only 3/16 inch deep.

Fig. 1 is a photomicrograph that shows considerable quantities of iron oxide in the crack. A portion of the insert was broken away with a hammer so as to expose the surfaces of the crack. Fig. 2, a photograph of the fracture, shows the oxidized condition of the crack. The fact that these surfaces were badly oxidized confirms the results of the metallographic examination.

Since there was no oxide coating on the outside surfaces of the insert as a result of the heat treatment, the very heavy deposit in the crack could not have been formed during the heat treatment. It is therefore concluded that the crack existed before the heat treatment.

A shallow, straight crack of this type with a heavy deposit of oxide is usually the result of a seam in the steel or a crack produced by improper cooling after the rolling operation. The failures are therefore considered to be the responsibility of the steel supplier, and entirely unconnected with the heat treatment.

Hardness readings taken on the insert showed the surfaces to be soft, with a Rockwell C hardness of only 61. The core, however, yielded a reading of 65, which is of course satisfactory. A case 0.004 inch deep was found on the surface during the metallographic examination, as shown in Fig. 3. The white area at the left side is retained austenite. Since the presence of austenite would account for the soft case, a chemical analysis of the case was made to investigate the presence of carbon, nitrogen, and nickel, which act as austenite stabilizers.

Chemical Analysis

Carbon and nickel determinations were made on millings taken to a depth of 0.030 inch below the surface. Before the millings were taken, the insert was softened: it was heated to 1550 F in a lead pot, held at this temperature for 2 hours, then slowly cooled to 800 F in the pot.

Notice that the description of methods, particularly the chemical analyses, are general rather than detailed. The writer assumes that his readers will believe that the laboratory used suitable methods. He assumes that they are interested primarily in the results of the tests rather than in the details of the methods used.

Notice that the next-to-last paragraph contains a qualification—a statement that it is impossible to tell what role the nickel played in producing the soft case.

The discussion ends with several conclusions that stem from preceding observations. Several other conclusions appeared earlier. All the important ones are restated in the summary.

The nitrogen content of the case was determined from a sliver 0.030 inch thick. The sliver was obtained by grinding away the core metal of the hardened insert on a surface grinder in order to avoid the possibility of decomposing any nitrides by a softening operation.

For comparison, carbon, nickel, and nitrogen analyses were repeated on drillings taken from the core metal. In addition, machinings taken to a depth of only 0.006 inch were analyzed for nickel. Results of the chemical analyses are given in the table below.

Partial Analysis of Cutter Insert

	% Nickel	% Carbon	% Nitrogen
Surface, 0.006 inch deep	0.34	----	----
Surface, 0.030 inch deep	0.108	0.77	0.47
Core	0.096	0.72	0.019

The increased content of nickel at the surface makes it evident that the insert was contaminated with nickel from the electrode of the Dunster salt bath. The increased quantities of carbon and nitrogen at the surface indicate that the salts used in the bath could not have been neutral.

Since nickel, carbon, and nitrogen all tend to stabilize austenite, it may be concluded that the use of this particular salt bath produced the soft case. Since the case contained excess carbon and nitrogen in addition to the nickel contamination, it is impossible to tell what role the nickel played in producing the soft case.

Since the crack was shown to have been present before the heat treatment, the increases in nickel, nitrogen, and carbon that took place during hardening could not have caused the crack.

The original report was Mimeographed. The illustrations were photographic prints pasted to the pages following the text. It would probably have been more helpful to the reader to put the photographs in suitable gaps in the text pages. But such an arrangement would have complicated production of the report, and in their present position the photographs are reasonably close to the related text.

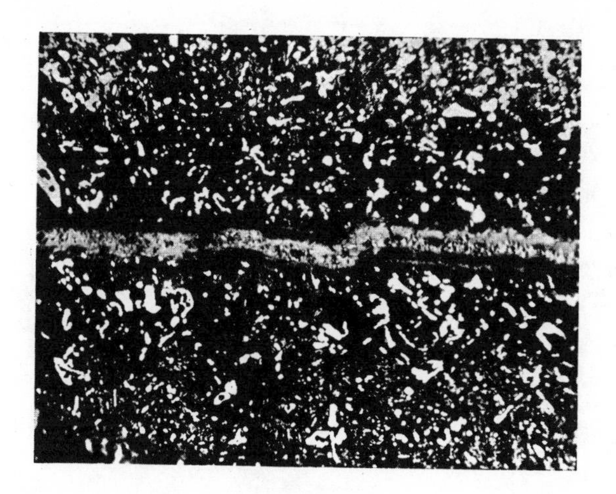

Fig. 1. Nital Etch 500x

Photomicrograph of sample taken midway along the crack, showing the deposit of iron oxide in the crack.

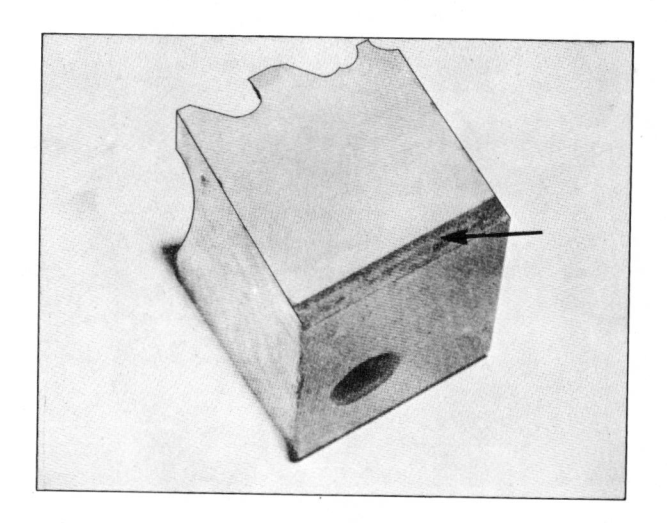

Fig. 2. Photograph of fractured portion of sample. The oxidized area (see arrow) is one surface of the crack.

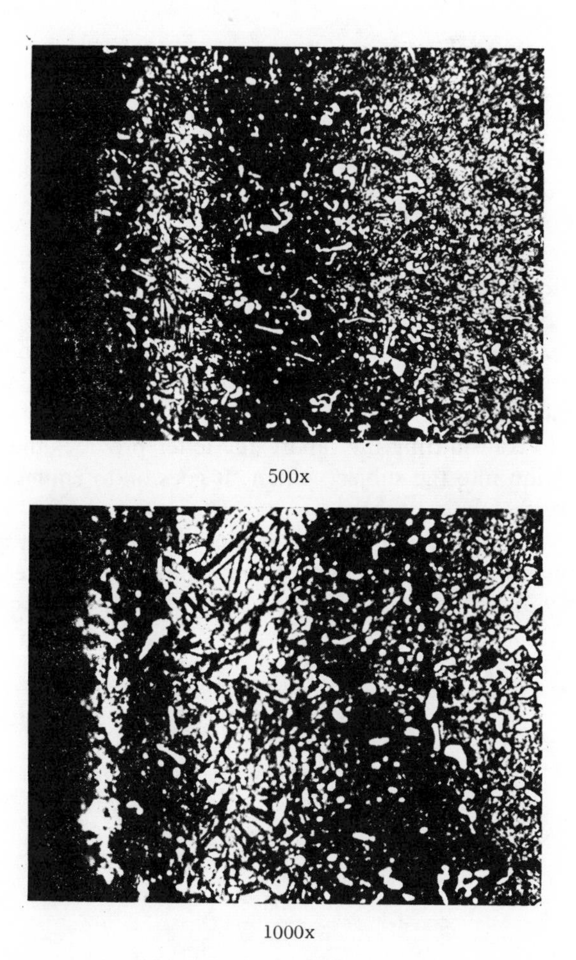

500x

1000x

Fig. 3. Nital etch. Photomicrographs taken at the edge of sample showing the presence of a case.

21.2.18 Specimen Letter of Transmittal

The specimen reproduced here illustrates some of the principles and functions of the letter of transmittal.

The survey referred to was obviously made on assignment. Under this circumstance, the writer undoubtedly felt that a letter of transmittal would be appropriate. He was able, therefore, to direct the report to a specific person within the organization.

In addition to transmitting the report, the letter provides the reader with a general orientation into the subject matter. It goes on to enumerate the objectives of the survey.

In the last paragraph, the letter makes a pertinent comment about the limits imposed on the survey. It closes by making a definite recommendation about future action.

21.2.18 SPECIMEN LETTER OF TRANSMITTAL

MARTIN ASSOCIATES, INC.
Management Consultants
San Francisco, California 94114

August 25, 1970

The Federal Reserve Bank of Santa Fe
20 Diamond Street
Santa Fe, New Mexico

Attention: Mr. George Sissell, First Vice President

Gentlemen: C-58496

We are submitting a report on a survey covering fields of industrial opportunity in the Southwest.

The survey was based on a review of the material and human resources of the region, and on technological advance. The objective was to discover new markets for products now made by Southwestern companies, methods for introducing newly developed products, and opportunities for establishing new industries suitable for the region. The emphasis has been placed principally on growth industries.

Because many of the staff members of Martin Associates, Inc., contributed technical analyses of products, the result has been a symposium covering a broad range of industrial processes and situations.

The purpose of the survey was only to identify manufacturing opportunities justifying further investigation. Interested organizations will, of course, require that those investigations be made before they commit themselves financially.

Respectfully submitted,

John J. Martin

John J. Martin

JJM:arw

21.2.19 Specimen Formal Instructions

This example of instructions—simple as it is—is in formal style. These instructions were written by an undergraduate engineering student.

21.2.19 SPECIMEN FORMAL INSTRUCTIONS

A WEEKEND PROJECT

BUILD A WOODEN BOOKCASE
WITHOUT NAILS

Here is a weekend project for building a simple and useful bookcase that will belong in any den or living room. You won't use one nail in this project. Pegs reinforce the joints and add a rustic appearance. The case can hold up to 100 books and costs less than $10.

TOOLS

- Router & guide
- Drill
- Chisel
- Hammer
- Handsaw
- Ruler and pencil

MATERIALS

Wood - clear pine

Piece Number (see diagram)	Dimension
a & b	34" x 3/4" x 8"
c & d	38" x 3/4" x 9-1/2"
e	36" x 1-1/2" x 10"
f	36" x 1-1/2" x 12"

Minwax stain

Minwax protector

Sandpaper

2

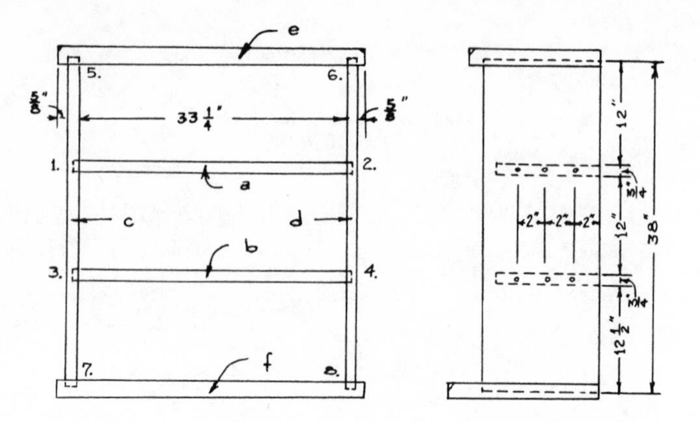

PROCEDURE:

1. Check the wood at the mill for warping or other flaws.

   Have the wood cut to proper size.

2. Hand-sand each piece to a uniform desired texture.

3. Lay off joints 1, 2, 3, & 4 on c & d with a pencil, using the dimensions in the diagram.

   Keeping within the pencil marks, rout out the joints to a 3/8" depth.

   Use a chisel to square the corners.

4. Check the joints by fitting the four pieces together. If necessary, widen the joints with the router, chisel, or sandpaper. Keep the joint snug.

5. Mark off 3 points 2" apart at each joint.

   Drill 3/8"-diameter holes for each peg at least 2" deep.

6. Lay off joints 5, 6, 7, & 8 on e & f and rout these to 5/8" depth.

   Chisel the corners. Pegs are not used at these joints.

3

7. Join all 6 pieces to check for mistakes.

   If e doesn't fit properly at the joints, widen the joints with the router in the needed direction to the point where c & d can just be forced into the joint.

8. Take the bookcase apart and lightly sand all pieces again and round all the forward edges.

   If a power jointer is available, cut the top corner of e & f at a 45° angle and at a depth of 1/4".

9. Wipe each piece clean with a damp cloth and apply a light coat of Minwax stain to all pieces. Let dry and repeat.

   Finally apply a coat of Minwax protector.

10. Cut 12 pegs, 2" long. Stain one end of each peg.

11. Assemble the bookcase and drive the pegs into the drill holes.

21.2.20 Specimen Formal Instructions

This example consists of the opening page from a relatively long and sophisticated industrial instruction manual.

THE HARRIS VITO MONITOR

Instructions for Turning On

The Harris Vito Monitor Model 780 is an automatic cardiac monitoring system, complete with its own built-in electrocardiograph and pacemaker, used to monitor the condition of a cardiac patient—or any other patient—either during diagnosis, in the operating room, or during recovery. It will not only monitor, but will also automatically pace the patient's heart in case of cardiac distress.

Necessary Supplies

Before you apply the Vito Monitor to the patient, you will need the following supplies:

1. A sturdy reliable table. This should be within 6 feet of a power socket: 115 volts, 60 cycles a-c.
2. ECG patient cable, electrodes, Redux electrode paste, and supplies—medical alcohol, sterile cotton, and adhesive tape.
3. Blood pressure plethysmograph and cable to connect it to the Vito Monitor.
4. Instructions on what kind of operation is required.

Turning on the Vito Monitor

1. Press the OFF button. This insures that the Vito Monitor will stay off until you are ready for it.
2. Check the warning on the rear of the case to see what frequencies should be used.
3. Unwrap the power cord from the rear and plug it firmly into a 115-volt 60-cycle a-c power socket.

CAUTION: You will see that the power cord has a 3-wire safety plug at the end. If you have only a 2-wire socket for the power outlet, use the 3-to-2 adapter supplied as a standard accessory. Connect the wire from the adapter to a good electrical ground, such as a cold-water pipe.

4. Press down the STOP button before turning on the Vito Monitor.
5. Turn the PACEMAKER control wheel to OFF.
6. Set the EXT-INT switch at the rear to INT so that pacing current will be at a minimum if it is accidentally turned on. Press down the STANDBY button to start the Vito Monitor. The power circuits will come on, the POWER and the INOPERATE lights will come on, but all other circuits will stay off.

At this point the Vito Monitor is connected to power and turned on to STANDBY, waiting for you to connect to the patient and to make the necessary adjustments prior to operating.

21.2.21 Specimen Proposal Letter

Here is an example of a proposal in letter form, written by students involved in a research project.

In this letter identification was easy to establish as the writer and other students mentioned in the letter were known to the reader. Once identification has been established, the writer describes the problem and gives the reason for writing the proposal.

As in most reports, the most important element—the proposal—is summarized step by step in as clear a way as possible.

Then follow the details, the logic behind the proposal, and other means of convincing the reader.

21.2.21 SPECIMEN PROPOSAL LETTER

Donald Sampson House
Pennsylvania Institute of Technology
October 1, 1969

Dr. D. M. Evans, Chairman
Department of Electrical Engineering
Pennsylvania Institute of Technology

Dear Dr. Evans:

Graduate students in the Systems Division of the department have been handicapped by not receiving clear information about computer programs already in existence and by a lack of methods for developing new programs. As a result, each new group of students must spend valuable time in going over the same material, time which they might better use on their research projects.

Proposal

For these reasons, it is proposed that:

1. A library of computer programs should be prepared within the department.

2. A manual of information should be written to accompany the programs.

3. A group of five or six graduate students should be permitted to develop and document the programs.

Details of Proposal

A number of computer programs are already in the possession of the Systems Division. However, they have never been classified and cataloged. The first phase of the procedure would be to put these programs in some definite form.

The next step would be to decide what programs are necessary and then to set about writing them, using those programs already in existence.

Since the main objective is to produce computer programs that can be used by the students and the faculty, the major part of the job would be to write clear and definite instructions.

In addition, a manual of information should be prepared showing how the programs were developed, what algorithms were used, and any limitations or possible problems. The manual should contain a complete flowchart of each part of the program. The manual would serve to standardize procedures.

The endorsements and a simple statement of costs complete the proposal except for the final paragraph, where the writer once more tries to persuade the reader that the proposal is worthwhile and then makes a request for direct action.

21.2.21 SPECIMEN PROPOSAL LETTER

Endorsement

Dr. John R. Fletcher, Systems Division, made the original suggestion leading to this proposal. He feels that the project could be handled by graduate students, as they work closely with computer programs.

Dr. E. H. Rutgers, professor of mathematics, endorses this proposal and is willing to serve as an advisor. Dr. Rutgers's principal contribution would come from his knowledge of numerical methods of computation.

A group of graduate students are enthusiastic over the idea and are willing to carry out the project; they have delegated the author as spokesman and probable supervisor.

Costs

The graduate students are willing to serve without pay. The only expenses foreseeable are those for necessary supplies, secretarial work, and computer time.

The graduate students hope that you will agree to this proposal. The project outlined should yield results that would be really useful to the Systems Division and perhaps to the rest of the professional community. We would appreciate it if you would let us know your response to this proposal within the next two weeks. Within a few months, greater demands will be made upon our time.

Sincerely,

Anton P. Holder

Anton P. Holder
Graduate Student, Systems Division

21.2.22 Specimen Formal Proposal

We have not reproduced here an entire formal proposal; only portions have been included for inspection.

This proposal was made to the United States Department of Commerce for support for a feasibility study. It concerns specific items of research with the purpose of determining whether a full-scale project would be justified.

As in a report, the title page carries the names of the authors—in this case, a faculty committee identified on the second page—together with the name of the proposed sponsor and file and contract numbers.

The preamble is similar to a foreword in a report. It gives some background information and suggests some applications of the vehicle being studied. It ends with a clear-cut statement of the scope of the proposed study.

We have included the complete table of contents to show the various subjects being discussed. However, only Area I of the study has been reproduced.

Area I, as do all the other sections, carries its own title page. This helps to separate it from the other sections and gives the name of the principal investigator.

Part A serves as a sample of the full-length proposal. It contains the background of this particular section, including statements on theory, and closes with the purpose. Notice that the section on method of approach contains items in parallel construction and that these items are numbered and listed. References to research already made are important in convincing the proposed sponsors that the study is feasible, as is a section on personnel.

Part A closes with a cost estimate. At the request of the authors, we have deleted the actual figures on cost. The other two parts follow an arrangement similar to that used in Part A.

To complete the proposal outline, we have included biographical information on the two men mentioned in Part A. These lists of accomplishments show how important authoritative personnel are to a proposal of this kind.

21.2.22 SPECIMEN FORMAL PROPOSAL

Rensselaer Polytechnic Institute

Proposal No. 42(66R)SE(1)

FEASIBILITY OF A NEW MODE OF MASS TRANSPORTATION

Submitted on behalf of

Faculty Committee on Transportation

of the

School of Engineering

to the

UNITED STATES DEPARTMENT OF COMMERCE

for supplemental funds and authorization

under Contract No. C-117-66(NEG)

Troy, New York

October 1965

ii

FACULTY COMMITTEE ON TRANSPORTATION

SCHOOL OF ENGINEERING

Russell M. Lewis (Chairman), Associate Professor of Civil Engineering

Dean N. Arden, Professor of Electrical Engineering

William B. Brower, Jr., Associate Professor of Aeronautical Engineering and Astronautics

Joseph V. Foa, Professor of Aeronautical Engineering

Imsong Lee, Associate Professor of Electrical Engineering

PREAMBLE

During the past several years there has been considerable effort at Rensselaer Polytechnic Institute toward the development of a new mode of ground transportation. The system studied is one in which a fluid-supported vehicle propels itself through a tube at speeds comparable to those of air transportation. The tube is not evacuated, and power requirements are minimized through the use of appropriate flow control devices in the tube and of novel means of propulsion and energy supply.*

It is envisioned that this proposed mode is ideally suited for the rapid and economical movement of people. The most favorable application might be for intermediate-length trips of about 25 to several hundred miles. It appears that speeds of 250 to 2000 miles per hour could be achieved. Among its several advantages would be inherent safety, adaptability to complete automation, and complete independence from atmospheric or surface conditions.

The Faculty Committee on Transportation of the School of Engineering proposes that it be authorized to conduct a feasibility study, confined to the technical evaluation of the critical components of the system. Four areas of research are crucial in establishing the technical feasibility of this new mode of transportation; therefore, this proposal is subdivided into the areas of propulsion, electrical power, stability and control, and a test facility.

---

*<u>High-Speed Mass Transportation - A New Concept</u>, Rensselaer Polytechnic Institute, July 1965.

CONTENTS

21.2.22 SPECIMEN FORMAL PROPOSAL

AREA I

THE PROPULSION OF VEHICLES
IN NONEVACUATED TUBES

by

J. V. Foa
Professor of Aeronautical Engineering

RENSSELAER POLYTECHNIC INSTITUTE

October 1965

2

INTRODUCTION

This area of research covers the propulsion of vehicles in nonevacuated tubes. PART A considers general technological feasibility. PART B explores feasibility of a specific mechanism -- namely, fluid-bladed propellers. PART C is essentially an addendum to PART B, covering work of a supporting nature that is not considered in the Cost Estimate.

PART A

STUDY OF THE FEASIBILITY OF INTERNAL MODES OF PROPULSION IN TUBES

BACKGROUND

The object of this part is to determine the technological feasibility of new modes of propulsion of fluid-supported vehicles in nonevacuated tubes.

Reference IA-1 describes a mode of thrust generation, called "internal," which promises important advantages of economy and safety in the propulsion of high-speed, fluid-supported vehicles in nonevacuated tubes. An "internally propelled" vehicle derives its thrust from the fore-to-aft transfer of the adjacent air masses within the tube. The transfer flow may be induced by a pusher or by a tractor propeller, or by a fan, or by an ejector pump, or even by heat addition alone, as in a ramjet.

The power required depends rather critically on the manner in which the transfer is effected (Ref. IA-2). In every case, the least power is required with the transfer mode that produces the least disturbances -- hence the least dissipation of energy -- in the air columns that occupy the remainder of the tube. The optimum transfer mode is not unique: it depends on the blockage ratio, on the shape of the vehicle, on the cruising speed, and on the component efficiencies of the transfer mechanism. The power required can be further reduced through control of the flow disturbances in the tube by means of gates, baffles, or valves.

A few specific methods of propulsion conforming to the basic scheme of a regulated air transfer have already been investigated. However, no meaningful comparison of these methods from the standpoint of power required has yet been possible, primarily because of unresolved questions concerning the nature of the flow in the wake of the vehicle (Refs. IA-2 and IA-3). Because of these uncertainties, unevenly conservative assumptions have had to be made in the analyses of the various modes.

The purpose of the proposed study is, in the first place, to develop general procedures and/or computational programs for the calculation of the power that is required to propel a fluid-supported vehicle in a tube, in relation to the size and aerodynamic characteristics of the vehicle, to its cruising speed, and to the mode of propulsion; in the second place, to apply these procedures to configurations and operating conditions of immediate interest, for the purpose of establishing the technological feasibility of internal modes of propulsion in tubes.

METHOD OF APPROACH

The proposed research will comprise the following steps:

1. Theoretical study of the flow in the wake of the vehicle, with special attention to the behavior of the flow when the tube is neither long enough to permit treatment of the flow as steady in the vehicle-fixed frame of reference nor so short that the dominant effect is that of reverberating pressure waves. Since these situations involve flows which are nonsteady in all frames of reference, they will be treated by the method of characteristics. The analysis will account for viscous effects and for heat exchanges with the surroundings.

4

2. Experimental verification and supplementation of theories developed under 1. The self-propelling vehicle in steady motion will be simulated in these experiments by very simple models. For example, a propagating deflagration wave could simulate a vehicle traveling subsonically, and a detonation wave could simulate a supersonic vehicle.

3. Analysis of flow losses and power requirements with various mechanisms of propulsion for vehicles traveling at constant speed in tubes of finite length. The propulsion mechanisms considered will include the purely internal modes (mechanical, thermal, and thermomechanical) and certain combinations of internal and external modes (e.g., propeller and linear induction motor drive).

4. Study of the dependence of the power required on interactions or controls applied to the flow outside the vehicle, at fixed locations within the tube.

5. Numerical applications of the theoretical results to situations of practical interest.

6. Engineering evaluation of the most practical methods.

REFERENCES

IA-1. Foa, J. V., "Propulsion," Section A of High-Speed Mass Transportation - A New Concept, Rensselaer Polytechnic Institute, July 1965.

IA-2. Foa, J. V., "Propulsion of a Vehicle in a Tube," Rensselaer Polytechnic Institute, Tech. Rept. TR AE 6404, June 1964.

IA-3. Hagerup, H. J., "A Note on the Flow Induced by a Disturbance Traveling in a Tube," Section H of Research on the Dynamics of a Vehicle in a Tube, Rensselaer Polytechnic Institute, Tech. Rept. TR AE 6406, June 1964.

PERSONNEL

The research will be directed by Dr. J. V. Foa. Participating personnel will include Dr. H. J. Hagerup, associate professor of aeronautical engineering, two Ph.D. candidates, one M.S. candidate, and one undergraduate student.

Biographical information on senior personnel is appended.

COST ESTIMATE - nine months

(AREA I - PART A)

Personnel

J. V. Foa
20% time - 6 academic months
35% time - 3 summer months
H. J. Hagerup
20% time - 6 academic months
35% time - 3 summer months
Graduate Assistant
1/2 time - 6 academic months
full time - 3 summer months
Machinist - 20% time, 9 months
Undergraduate assistance - as required
Typist - as required

Materials, Supplies, and Services

Expendable Experimental Supplies
Miscellaneous

Equipment

Kistler Gauge Calibrator
Welch 1400-B Vacuum Pump
Trigger Amplifiers
Miscellaneous

Reports and Publications

Travel and Communication

Three trips to Washington
Conferences with other researchers in the field

Indirect Cost

51.87% of Total Personnel Payments

NOTE: This Cost Estimate is summarized in, and is hereby made a part of, the attached SUMMARY COST ESTIMATE.

6

PART B

RESEARCH ON BLADELESS FANS AND PROPELLERS
FOR THE PROPULSION OF HIGH-SPEED VEHICLES IN TUBES

INTRODUCTION

The purpose of this part is to explore the feasibility of propelling vehicles through tubes by means of fluid-bladed propellers. These devices would have very important advantages from the standpoints of weight, safety, and adaptability to the operating conditions of a fluid-supported vehicle in a tube.

A land vehicle traveling at very high speed must be supported from all sides, but must also be left free to tilt itself to the right angle of bank in every turn. These requirements, as well as all considerations of safety, confort, guidance, communication, control, and economy, call for the use of large-clearance fluid suspension devices, with a tube as the guideway.

A system so designed is discussed in Reference IB-1. Section A of this reference deals, in particular, with certain methods of flow induction that can be used to propel a fluid-supported vehicle in a tube. One of the difficulties here is the large clearance that must be provided between the tube wall and all mechanical components of the thrust generator, because of the freedom of transverse motion that the vehicle is allowed by its suspension.

A method of flow induction that appears to be particularly well suited for this application is that of "cryptosteady pressure exchange," which has been under study at Rensselaer for several years (Refs. IB-2 through IB-7) and has recently been receiving attention elsewhere as well (Refs. IB-8 through IB-11). In cryptosteady pressure exchange, energy is transferred nondissipatively from one flow to another by moving pressure fields. The process is, therefore, nonsteady. It admits, however, a frame of reference in which it is steady. It can, therefore, be analyzed and controlled as a steady-flow process in this unique frame of reference, while retaining all the potential advantages of nonsteady-flow processes in the frame of reference in which it is utilized. The designation "cryptosteady" serves to distinguish this process from those nonsteady exchanges that are carried out by propagating pressure waves in conventional pressure exchangers, in shock tubes, and in intermittent jets.

In a simple embodiment of this concept, the frame of reference in which the interacting flows are steady rotates at a constant angular velocity about the axis of the vehicle. A flow of high-pressure air from a compressor (primary fluid) is made to issue through slanted orifices on the periphery of a rotating member which is driven by the reaction of the issuing jets themselves. At every instant, the primary fluid which has emerged during a brief and immediately preceding time interval from each rotating orifice occupies a spiral or helical region in space, which rotates about the same axis and at the same angular velocity. Therefore the interacting flows exchange energy in the frame of reference of the vehicle by a mechanism essentially similar to that of turbopumps or turbopropellers, although the "blades" are now patterns rather than bodies of abiding material. In a variation of this embodiment, rotating pseudoblades are generated without any moving mechanical part, through controlled rotating stall in stationary cascades (Ref. IB-12).

Experimental results so far have confirmed the remarkable potential of this mode of propulsion but have been erratic. Much uncertainty still exists about the details of the interaction. As a result, rational design criteria still remain to be established. There is little doubt, however, that the successful development of a cryptosteady pressure exchanger for propulsion would provide the simplest, safest, and most economical solution to the problem of propelling a fluid-supported high-speed vehicle in a tube. As noted in Section A of Reference IB-1, the blades of such a device would be free from structural limitations and would always be in sealing contact with the boundaries of the transfer passage, despite the freedom of transverse motion of the vehicle relative to the tube.

METHOD OF APPROACH

The proposed program of research includes the following:

1. Extension of existing theories to cover
   a. the effect of finite thickness of the primary stream,
   b. effects of energy exchanges following the nondissipative phase,
   c. effects of mixing during the deflection phase.

2. Experiments aimed at an understanding of the mechanisms of disintegration of the driving streams (pseudoblades). Fluid pseudoblades will be produced under controlled conditions and visualized by smoke or dye as they penetrate at various "collision angles" into a transverse stream. Particular attention will be given to the relation between the initial turbulence level of the pseudoblades and their penetration distance, for various values of pertinent nondimensional parameters (Reynolds number, stream velocity ratio, collision angle).

3. Experimental determination of the energy transfer efficiencies of cryptosteady energy exchanges designed on the basis of the studies outlined above.

REFERENCES

IB-1. High-Speed Mass Transportation - A New Concept, Rensselaer Polytechnic Institute, July 1965.

IB-2. Foa, J. V., "A New Method of Energy Exchange Between Flows and Some of Its Applications," Rensselaer Polytechnic Institute, Tech. Rept. TR AE 5509, December 1955.

IB-3. Foa, J. V., "Cryptosteady Energy Exchange," Rensselaer Polytechnic Institute, Tech. Rept. TR AE 6202, March 1962.

IB-4. Guman, W. J., "Exploratory Experimental Study of a New Method of Energy Exchange Between Steady Flows," Rensselaer Polytechnic Institute, Tech. Rept. TR AE 5811, May 1958.

IB-5. Vennos, S. L. N., "Experiments with Water-Water Bladeless Propellers," Rensselaer Polytechnic Institute, Tech. Rept. TR AE 6106, May 1961.

IB-6. Foa, J. V., "A Method of Energy Exchange," ARS Jour., 32, No. 9, 1396-1398, September 1962.

8

IB-7. Foa, J. V., "A Vaneless Turbopump," AIAA Jour., 1, No. 2, 466-467, February 1963.

IB-8. Hohenemser, K. H., "Preliminary Analysis of a New Type of Thrust Augmenter," Proceedings, the Fourth National Congress of Applied Mechanics, Berkeley, June 1962.

IB-9. McNair, C. H., Jr., M. S. Thesis, Georgia Institute of Technology, June 1963.

IB-10. Waters, M. H., "Cryptosteady Pressure Exchange," U.S. Naval Air Turbine Test Station, Trenton, New Jersey, Tech. Note NATTS-ATL-TN 15, July 1962.

IB-11. Hohenemser, K. H., "Flow Induction by Rotary Jets," McDonnel Aircraft Corporation, Rept. No. B353, 30 December 1964.

IB-12. U.S. Pat. Appl'n. Ser. No. 213,560, filed 30 July 1962, assigned to the U.S. Air Force.

PERSONNEL

The scientific personnel for the program will include Dr. J. V. Foa, principal investigator; Dr. W. B. Brower, Jr.; one graduate assistant; and one undergraduate assistant.

Biographical information on senior personnel is appended.

COST ESTIMATE - nine months

(AREA I - PART B)

Personnel

J. V. Foa
20% time - 6 academic months
35% time - 3 summer months
W. B. Brower
10% time - 6 academic months
25% time - 3 summer months
Graduate Assistant
1/2 time - 6 academic months
full time - 3 summer months
Machinist - 25% time, 9 months
Undergraduate assistance - as required
Typist - as required

Materials, Services, and Supplies

Rotor Models
Expendable Laboratory Supplies
Modification of Existing Facilities
Miscellaneous

Reports and Publications

Travel and Communication

Conferences with other researchers in the same or related fields (Washington University in St. Louis, foreign research institutes)

Indirect Cost

51.87% of Total Personnel Payments

NOTE: This Cost Estimate is summarized in, and is hereby made a part of, the attached SUMMARY COST ESTIMATE.

PART C

THE GENERATION OF FLUID BLADES
THROUGH CONTROLLED ROTATING
STALL IN STATIONARY CASCADES

A modest incremental amount will probably be requested in the near future for an addendum to PART B of AREA I. This addendum will concern a study of the feasibility of generating a rotating pattern of fluid blades for bladeless fans or propellers through the excitation and control of rotating stall in stationary cascades, i.e., without any moving mechanical part.*

It is proposed that the performance of this additional task, if approved, be subcontracted to the Institute of Applied Mechanics and Aerodynamics of the Politecnico di Torino (Turin Institute of Technology, Italy), where this problem has recently been studied by Professor Carlo Grillo Pasquarelli, with the part-time collaboration of Professor Fiorenzo Quori. Professor Gillo Pasquarelli's study, being unsponsored, had to be discontinued in the spring of this year because basic experiments essential to its meaningful continuation could not be conducted for lack of funds.

Because of his competence, demonstrated interest, and recent activity in the study of controlled rotating stall, Professor Grillo Pasquarelli is believed to be the man presently best qualified to examine and answer the question of feasibility of this process within the short time span of this first phase.

---

*Foa, J. V., "Cryptosteady Energy Exchange," Rensselaer Polytechnic Institute, Tech. Rept. TR AE 6202, March 1962.

21.2.22 SPECIMEN FORMAL PROPOSAL

BIOGRAPHICAL INFORMATION

Joseph V. Foa

Professor of Aeronautical Engineering
Chairman Department of Aeronautical Engineering and Astronautics
Rensselaer Polytechnic Institute

Education

Dr. Mech. Eng.	Politecnico di Torino, Italy, 1931
Dr. Aero. Eng.	University of Rome, 1933

Professional Experience

1933-35	Research and Project Engineer, Piaggio Aircraft Corporation, Italy
1936-37	Chief Engineer, Caproni Aircraft Corporation, Italy
1937-39	Project Engineer and Consultant, Piaggio Aircraft Corporation, Italy
1939-40	Project Engineer, Bellanca Aircraft Corporation, New Castle, Delaware
1940-42	Instructor of Aircraft Structures and Design, University of Minnesota
1942	Consulting Engineer
1943-46	Head of Aeronautical Design Section, Research Laboratory, Curtiss-Wright Corporation, Buffalo, New York
1946-52	Head of Propulsion Branch, Cornell Aeronautical Laboratory, Buffalo, New York
1952-date	Member of Faculty, Rensselaer Polytechnic Institute
	Professional Engineer's License, State of New York, No. 38835 PE

Important Consulting Work since 1952

Cornell Aeronautical Laboratory
General Electric Company
General Motors Research Laboratory
Martin Company
Grumman Aircraft Corporation
Flettner Aircraft Corporation
Battelle Memorial Institute
McDonnell Aircraft Corporation
Pratt and Whitney Aircraft Corporation
U.S. Army Research Office-Durham
National Academy of Sciences

Society Memberships

American Institute of Aeronautics and Astronautics - Associate Fellow
Sigma Gamma Tau (national honorary aeronautical engineering society)
Society of the Sigma Xi

Joseph V. Foa

Patents

New type of aileron:
Italian patent No. 347,173 (1937)
French patent No. 831,139 (1938)

Method of energy exchange:
British patent No. 860,073 (1957)
Canadian patent No. 622,784 (1961)
United States patent No. 3,046,732 (1962)

Honors and Awards

Naval Ordnance Development Award (for exceptional service to Naval Ordnance development), December 1945.
Army certificate of appreciation for services on the Technical Industrial Intelligence Committee of the Joint Chiefs of Staff during World War II, May 1951.

Publications

"Single-Flow Jet Engines - A Generalized Treatment," Journal of the American Rocket Society, 21, 115-126 (1951).

"Propulsive Ducts at the Tips of Rotor Blades," Journal of the Aeronautical Sciences, 23, No. 4 (1956) (with A. Gail and T. R. Goodman).

"Pressure Exchange," feature article, Applied Mechanics Reviews, 11, No. 12 (1958).

"Intermittent Jets," Section XII-F of High-Speed Aerodynamics and Jet Propulsion, Princeton Unitversity Press, 1959.

"Elements of Flight Propulsion," John Wiley & Sons, Inc., New York, 1960.

"Land Transport at Air Transport Speeds," National Defense Transportation Journal, 18, No. 4 (July-August 1962).

"A Method of Energy Exchange," American Rocket Society Journal, 32, No. 9 (September 1962).

"A Vaneless Turbopump," Rensselaer Polytechnic Institute, Technical Report No. TR AE 6209, 1962; American Institute of Aeronautics and Astronautics Journal, 1, No. 2, 466-467 (February 1963).

"Transportation," in section "Research Frontiers," Saturday Review, 7 November 1964.

Henrik J. Hagerup

Associate Professor of Aeronautical Engineering
Rensselaer Polytechnic Institute

Education

B.S.	Massachusetts Institute of Technology, 1955
M.S.	Massachusetts Institute of Technology, 1956
Ph.D.	Princeton University, 1963

Professional Career

1962-date	Member of Faculty, Rensselaer Polytechnic Institute

Society Memberships

American Institute of Aeronautics and Astronautics
Sigma Gamma Tau
Tau Beta Pi
Society of the Sigma Xi
American Association for the Advancement of Science

Publications

Equilibrium Configurations of Shock Waves in Curved Channels, Master's Thesis, Massachusetts Institute of Technology, 1956.

Linearized Viscous Flow Near Surface Discontinuities, Doctoral Thesis, Princeton University, 1963.

Note on the Flow Induced by a Disturbance Traveling in a Tube. Part One - Stationary Flow, TR AE 6405, Department of Aeronautical Engineering and Astronautics, Rensselaer Polytechnic Institute, 1964.

Incompressible Cavity Flows in Two Dimensions at Low Reynolds Number, Report 691, Gas Dynamics Laboratory, Princeton University, 1964.

21.2.23 Specimen Technical Article

This article was written by a student with a background in chemistry. It reflects her specialization and her scientific interest. It was published in an engineering college magazine.

The lead paragraph introduces the subject quickly and at the same time is definite and specific through the mention of a date, a symposium, and two well-known scientists.

The second paragraph introduces the unifying idea implicit in the title: that the article is to define the chemical laser and show its historical development.

Although it is technical, this article maintains a concrete and specific level. Read the article carefully; you will see that the author uses the two principles of illustration and interpretation. She builds definition by describing processes and experiments. She also uses a step-by-step analysis as in the section on pumping energy.

The article ends in a conventional, yet effective, manner with the uses of the laser and a statement on the future. A more detailed analysis of this article will reveal many of the devices discussed in Sec. 12.3.2.

21.2.23 SPECIMEN TECHNICAL ARTICLE

CHEMICAL LASERS

by Mary Albert Porter

In 1964 a symposium was held in California to discuss the possibilities of chemical lasers. Although no operative systems were reported, valuable theoretical discussions predicted their existence. When Jerome V. V. Kasper and George C. Pimental of the University of California at Berkeley reported the successful operation of photodissociation and explosion lasers early in 1965, the pathway for a new and exciting laser technology was begun.

A study of the discovery and operation of the chemical laser reveals the promises it offers to this new industry. Its development was preceded in 1960 by the earliest successful laser—a pulsed ruby laser operated by Theodore H. Maiman, then of the Hughes Research Laboratories. This was followed by other lasers made of gas, solids, or liquids and powered by electricity, light, or radio waves. The chemical laser developed as scientists predicted greater laser efficiency and broader wavelengths of light radiation.

L.A.S.E.R.

The operation of the laser depends upon the theory discovered by Planck in 1900 and expanded by Einstein seventeen years later: An atom prefers to remain in its ground, or lowest energy, state. Excited by electromagnetic radiation (light), the atom absorbs a photon, a quantum of light. The excited atom is in an unstable or higher energy state and tends to spontaneously emit this photon. If, however, the excited atom is irradiated with a photon having energy identical to that spontaneously emitted, two photons are released, and stimulated emission occurs. As the light moves through the medium, each quantum of light is amplified as stimulated emission of other quanta are induced. Hence, the acronym: *L*ight *A*mplification by *S*timulated *E*mission of *R*adiation.

Scientists designing lasers have thus been confronted with a critical problem: how to make stimulated emission predominate over spontaneous emission and absorption. They must produce an inverted population of excited atoms. In Maiman's laser, a flash-tube light provided the energy necessary to raise the ruby atoms to the excited state. Once this is done, the photons are amplified until they reach a certain energy—a cascade of photons is suddenly released, producing an intense beam of coherent light at virtually a single wavelength.

The power or "pumping" source for Maiman's ruby laser was light. The active medium was the ruby crystal. Since the energy for pumping required external electric power and was inefficient, scientists became intrigued with the possibilities of an active medium that could pump itself.

A chemical reaction releases its own energy through processes of chemical bond-breaking and bond-making. Once initiated, a chemical reaction could provide its own energy for pumping. Some chemical reactions were already known to produce population inversions. Theoretically, such a self-pumping system could be extremely efficient!

Chemical Laser

A chemical laser may be defined as a chemical system in which the energy distributed among the reaction products causes an excited-state population inversion and laser action. An extension of this definition considers an external energy source, such as electricity or light, needed to initiate the reaction. The main point of this extended definition is the making or breaking of a chemical bond and not the origin of energy required for these processes.

The photodissociation laser discovered by Kasper and Pimental is based on a photon-initiated chemical reaction that produces excited iodine atoms that release photons. The reaction is not self-sustaining, but the population inversion is caused by energy distribution caused by chemical bond-breaking.

C. K. N. Patel and his co-workers at Bell Telephone Laboratories have obtained laser emission in the infrared by passing an electric discharge through carbon monoxide and carbon dioxide at low pressures. Although the electric discharge is the main pumping source, final energy may be provided by chemical bond-making.

Bond-making and -breaking are also evident in the water-vapor laser developed by W. J. Witteman and R. Bleekrode at Philips Research Laboratory (Eindhoven, the Netherlands). Laser action was achieved in the far infrared. An electric discharge fragments the water molecule into hydrogen atoms and rotationally excited hydroxide radicals.

Explosion Laser

Amidst the development of many other chemical lasers, Kasper and Pimental's second discovery—the hydrogen and chlorine "explosion" laser—merits special attention for two reasons. As the first operating laser excited by a chemical reaction of two different substances, it has been the model for similar systems. Secondly, it demonstrates a new method for determining energy distributions in chemical reactions.

Examining any laser system, one finds that it has three main parts: a set of energy levels in an active material, a pumping system, and an optical cavity. The optical cavity augments the amplification of coherent light as mirrors on either end of this long tube reflect the light back and forth. When these photons trigger more energy from the molecules in the material, more photons are added to the horizontal ray. After the photons reach a certain energy level, they burst through the weaker mirror in a sharp beam of coherent light.

The Kasper-Pimental laser is a quartz tube containing one part chlorine and two parts hydrogen at reduced pressure. After the reaction is initiated by a photolytic flash, the gases "explode"—that is, react to form hydrogen chloride (HCl) and atomic chlorine (Cl). As the vibrationally excited HCl falls to a less energetic state, photons are released and laser radiation is emitted in the infrared.

Pumping Energy

In this system, the set of energy levels (chlorine and hydrogen gases) intrinsically supply the pumping energy once the reaction is initiated. The reactions for this system are in four steps: (1) The chlorine molecule is separated into two chlorine atoms by flash photolysis. (2) This chlorine atom then reacts with a hydrogen molecule to form hydrogen chloride (HCl) and a hydrogen atom. (3) The hydrogen atom reacts with molecular chlorine to form excited hydrogen chloride (HCl *) and a chlorine atom which continues step 2. (4) The excited hydrogen chloride emits a photon through stimulated emission.

Step 3 is the vital reaction for this process—the pumping reaction. It is an exothermic reaction, which means that heat is given off by the system. The net energy release is the difference between energies required for the bond-breaking of H_2 and the bond-making of HCl.

Excitation

An examination of the excited products of any reaction reveals that their internal energy is contained in four types of excitation: electronic, vibrational, rotational, and translational. Each of these types requires a certain amount of energy and results in a certain kind of radiation.

For example, electronic excitation requires over 100,000 calories per mole and emits light at visible or ultraviolet frequencies. Translational excitation, on the other hand, requires very little energy and is associated with heat. Vibrational excitation requires less energy than electronic excitation and is emitted in the infrared. Rotational excitation requires even less energy and emits in the microwave region of the spectrum.

Nonequilibrium excitation is a necessary condition for laser action. As the chemical system approaches equilibrium, however, the energy is redistributed among those four types of excitation. Since electronic and vibrational de-excitation are slower than all the other processes, not all the energy is immediately released as heat. Ideally, only electronic or vibrational excited states might result in an energy level system after a chemical reaction. If this occurred, there would be no lower-level population until de-excitation began—a chemical laser could be very efficient!

In the laser with the excited HCl molecules, however, the situation is far from the ideal. Electronic excitation is never attained, since there is not enough energy in the system. Besides energy released as heat, the remainder is stored as vibrational excitation and rotational and translational motion. This vibrational excitation facilitates stimulated emission and laser action in HCl *. In contrast, the Kaper-Pimental photodissociation laser that used iodine exhibited a very intense laser emission. The high gain in energy lends optimism to the use of the chemical laser for high power levels.

Disadvantages

Of course there are many problems with the chemical laser at this early stage of development. The way the reaction is initiated may impose limiting factors. Vibrational excitation lasers may exhibit reduced efficiency because of occupation of rotationally excited sublevels. Moreover, lack of knowledge about chemical reactions applicable to lasers impedes progress on new systems.

This last limitation may actually provide a realistic use for the chemical laser. At this time, little is known about very basic theory of chemical reactions. If the energetic states of products can be uncovered, further information on the distribution of energy in elementary reactions will be facilitated.

Chemical Uses of the Laser

Dr. Pimental illustrates the application of chemical lasers to kinetic studies in a preprint, *Chemical Information from Chemical Lasers.* In a chemical reaction, the overall rate constant, k, is the sum of various rate constants (k_0, k_1, k_2, etc.) of the excited states of the product molecules. Using chemical lasers to indirectly measure such smaller rate constants, Dr. Pimental demonstrated that detailed knowledge of chemical processes can be obtained.

Five flash-initiated chemical systems analogous to the hydrogen-chlorine explosion laser are applicable to this study. Further investigation of the hydrogen-chlorine laser enables identification of emitting laser transitions. Factors which influence laser gain (intensity) can also be chosen. Since laser operation occurs rapidly compared to deactivation of the vibrationally excited HCl * molecules, factors increasing the reaction rate of the pumping reaction increase laser efficiency.

Future

The future of the chemical laser is difficult to predict. Problems such as nonspecificity of chemical pumping and the resulting broad energy distribution of the product species over the electronic, vibrational, rotational, and translational levels are countered by other advantages. The large gain in intensity demonstrated by the Kasper-Pimental iodine laser may prove to be an asset of the chemical laser.

The application of the chemical laser reveals more about elementary chemical processes. It seems very probable that the chemical laser, still early in its development, will achieve successful and useful operations.

21.3 Bibliography

Dictionaries

The American College Dictionary. New York: Random House, Inc., 1965.

The American Heritage Dictionary of the English Language. Boston: Houghton Mifflin Company, 1970.

New College Standard Dictionary. New York: Funk & Wagnalls Company, 1969.

Concise Oxford Dictionary of Current English (5th ed.). New York: Oxford University Press, 1964.

The Random House Dictionary of the English Language: The College Edition. New York: Random House, Inc., 1968.

Rodale, J. I. *The Synonym Finder*. Emmaus, Pa.: Rodale Books, Inc., 1965

Tweney, C. F., and Hughes, L. E. C. *Chambers's Technical Dictionary* (3d ed.). New York: The Macmillan Company, 1958; no longer in print.

Webster's Seventh New Collegiate Dictionary. Springfield, Mass.: G. and C. Merriam Company, 1965.

Webster's Dictionary of Synonyms. Springfield, Mass.: G. and C. Merriam Company, 1967.

Webster's New World Dictionary. New York: Harcourt, Brace & World, Inc., 1964.

Handbooks and Textbooks

Brown, Clarence A., and Zoellner, Robert. *The Strategy of Composition*. New York: The Ronald Press Company, 1968.

Brown, Leland. *Communicating Facts and Ideas in Business*. Englewood Cliffs, N.J.: Prentice-Hall, Inc., 1961.

Comer, D. B., and Spielman, R. R. *Modern Technical and Industrial Reports*. New York: G. P. Putnam's Sons, 1962.

Connolly, Francis, and Levin, Gerald. *A Rhetoric Case Book* (3d ed.). New York: Harcourt, Brace & World, Inc., 1968.

Corbett, Edward P. J. *Classical Rhetoric for the Modern Student.* New York: Oxford University Press, 1965.

Damerst, William. *Resourceful Business Communication.* New York: Harcourt, Brace & World, Inc., 1966.

Daniel, Robert W. *A Contemporary Rhetoric.* Boston: Little, Brown & Company, 1967.

Davidson, Donald. *American Composition and Rhetoric* (5th ed.). New York: Charles Scribner's Sons, 1964.

Estrin, Herman A. (ed.). *Technical and Professional Writing; A Practical Anthology*. New York: Harcourt, Brace & World, Inc., 1963.

Evans, Bergen, and Evans, Cornelia. *A Dictionary of Contemporary American Usage.* New York: Random House, Inc., 1957.

Fowler, H. W. *A Dictionary of Modern English Usage* (2d ed.). New York: Oxford University Press, 1965.

Gunning, Robert. *New Guide to More Effective Writing in Business and Industry*. Boston: Industrial Education Institute, 1962.

Halverson, John, and Cooley, Mason. *Principles of Writing.* New York: The Macmillan Company, 1965.

Hodges, John C., and Whitten, Mary E. *Harbrace College Handbook* (6th ed.). New York: Harcourt, Brace & World, Inc., 1967.

Janis, J. Harold. *Writing and Communication in Business.* New York: The Macmillan Company, 1964.

Kapp, Reginald O. *The Presentation of Technical Information.* New York: The Macmillan Company, 1957.

McCrimmon, James M. *Writing with a Purpose* (4th ed.). Boston: Houghton Mifflin Company, 1967.

Mills, Gordon H., and Walter, John A. *Technical Writing* (rev. ed.). New York: Holt, Rinehart and Winston, Inc., 1962.

Mitchell, John H. *Handbook of Technical Communication.* Belmont, Calif.: Wadsworth Publishing Co., Inc., 1962.

——— *Writing for Technical and Professional Journals.* New York: John Wiley & Sons, Inc., 1968.

Needleman, Morriss H. *Handbook for Practical Composition.* New York: McGraw-Hill Book Company, Inc., 1968.

Pearsall, Thomas. *Audience Analysis for Technical Writing.* Beverly Hills, Calif.: Glencoe Press, 1969.

Rathbone, Robert R. *Communicating Technical Information.* Reading, Mass.: Addison-Wesley Publishing Company, Inc., 1966.

Rubenstein, S. Leonard, and Weaver, Robert G. *The Plain Rhetoric* (2d ed.). Rockleigh, N.J.: Allyn and Bacon, Inc., 1968.

Strunk, William, Jr., and White, E. B. *The Elements of Style*. New York: The Macmillan Company, 1959.

Tichy, Henrietta J. *Effective Writing for Engineers, Managers, Scientists*. New York: John Wiley & Sons, Inc., 1966.

University of Chicago Press. *A Manual of Style* (12th ed.). Chicago, 1969.

Weaver, Richard M. *Rhetoric and Composition* (revised with Richard S. Beal). New York: Holt, Rinehart and Winston, Inc., 1967.

Weisman, Herman M. *Technical Correspondence*. New York: John Wiley & Sons, Inc., 1968.

——— *Technical Report Writing*. Columbus, Ohio: Charles E. Merrill Books, Inc., 1966.

Wyld, Lionel D. *Preparing Effective Reports*. New York: The Odyssey Press, Inc., 1967.

Editorial Style Guide (2d ed.). Whippany, New Jersey: Bell Telephone Laboratories, 1967.

Technical Society Publications

American Chemical Society. *Handbook for Authors* (1st ed.). Washington, 1967.

American Institute of Electrical and Electronics Engineers. *Information for IEEE Authors*. New York, 1948; supplement, 1965.

American National Standards Institute. *Abbreviations for Scientific and Engineering Terms* (Z10.1–1941). New York, 1941.

——— *Abbreviations for Use on Drawings* (32.13–1950). New York, 1950.

Engineers' Council for Professional Development. *Speaking Can Be Easy . . . for Engineers too*. New York, 1950.

Articles

Britton, W. E. "What to Do About Hard Words." *STWP Review,* October 1964.

Cohen, Gerald I. "A World Without People." *IEEE Transactions on Engineering Writing and Speech,* Vol. ESW–12, No. 3, October 1969.

Gould, J. R. "Barriers to Effective Communication." *Journal of Business Communication,* Winter 1968.

——— "Psychology in Your Communications." *Chemical Engineering,* Dec. 5, 1966.

Kaysing, W. C. "Engineers and the Technical Writer." *Chemical Engineering,* July 18, 1966.

Kilpatrick, J. N. "Writing and Presenting the Technical Paper." *Naval Engineering Journal,* August 1965.

Koff, R. M. "Eight Steps to Better Engineering Writing." *Product Engineering,* March 15 through May 24, 1965.

Koral, R. L. "How to Write Technical Articles for Publication." *Air Conditioning, Heating, and Ventilation,* August 1968.

Law, J. W., and Braunstein, J. "Preparation of a Technical Article." *Bulletin of the American Society of Petroleum Geologists,* November 1964.

Miller, H. M. "A B C's of Good Technical Writing." *Supervisory Management,* August 1965.

Morris, S. "Engineering Proposals." *Machine Design,* May 13, 1965.

Quiller-Couch, Sir Arthur Thomas. "On Jargon," from *On the Art of Writing.* New York: G. P. Putnam's Sons, 1950.

Richards, R. W. "A New Look at Engineering Reports." *Civil Engineering,* October 1966.

Seidman, I. M. "Improving Engineering Reports and Talks." *IEEE Transactions on Engineering Writing and Speech,* July 1967.

Wainwright, G. "Effective Report Writing." *Aircraft Engineering,* April 1968.

Woodkins, O. J., and King, R. E. "Writing Persuasive Proposals," *Electronic Index,* February 1965.

Index